Neil Postman

Technopoly

Neil Postman was a critic, communications theorist, and Chair of the Department of Communication Arts and Sciences at New York University. In 1987 he was given the George Orwell Award for Clarity in Language by the National Council of Teachers of English. In 1989 he received the Distinguished Professor Award at New York University. In the spring of 1991 he was Laurence Lombard Visiting Professor of the Press and Public Policy at Harvard University. For ten years he was editor of *Et Cetera,* the journal of General Semantics. His seventeen previous books include *Teaching as a Subversive Activity* (with Charles Weingartner), *The Disappearance of Childhood, Amusing Ourselves to Death, Building a Bridge to the 18th Century,* and *Conscientious Objections*. He died in 2003.

ALSO BY NEIL POSTMAN

Building a Bridge to the 18th Century

The End of Education

Conscientious Objections

Teaching as a Subversive Activity
(with Charles Weingartner)

Crazy Talk, Stupid Talk

Teaching as a Conserving Activity

The Disappearance of Childhood

Amusing Ourselves to Death

Technopoly

Neil Postman

Technopoly

The Surrender of Culture to Technology

Vintage Books
A Division of Random House, Inc.
New York

FIRST VINTAGE BOOKS EDITION, APRIL 1993

 Published in the United States by Vintage Books, a division of Random House, Inc., New York, and simultaneously in Canada by Random House of Canada Limited, Toronto. Originally published in hardcover by Alfred A. Knopf, Inc., New York, in 1992.

Library of Congress Cataloging-in-Publication Data
Postman, Neil.
Technopoly: the surrender of culture to technology / Neil Postman.
p. cm.
Originally published: 1st ed. New York: Knopf, 1992.
Includes bibliographical references and index.
ISBN 0-679-74540-8 (pbk.)
1. Technology—Social aspects. I. Title.
T14.5.P667 1993
303.48'3—dc20 92-50584
CIP

Book design by Mia Vander Els

Manufactured in the United States of America

79C86

For Faye and Manny

Whether or not it draws on new scientific research,
technology is a branch of moral philosophy, not of science.

PAUL GOODMAN, *New Reformation*

Contents

Introduction

In 1959, Sir Charles Snow published *The Two Cultures and the Scientific Revolution,* which was both the title and the subject of the Rede Lecture he had given earlier at Cambridge University. The lecture was intended to illuminate what Sir Charles saw as a great problem of our age—the opposition of art and science, or, more precisely, the implacable hostility between literary intellectuals (sometimes called humanists) and physical scientists. The publication of the book caused a small rumble among academics (let us say, a 2.3 on the Richter Scale), not least because Snow came down so firmly on the side of the scientists, giving humanists ample reason and openings for sharp, funny, and nasty ripostes. But the controversy did not last long, and the book quickly faded from view. For good reason. Sir Charles had posed the wrong question, given the wrong argument, and therefore offered an irrelevant answer. Humanists and scientists have no quarrel, at least none that is of sufficient interest to most people.

Nonetheless, to Snow must go some considerable credit for noticing that there *are* two cultures, that they are in fierce opposition to each other, and that it is necessary for a great debate to ensue about the matter. Had he been attending less to the arcane dissatisfactions of those who dwell in faculty clubs and more to the lives of those who have never been in one, he would surely have seen that the argument is not between humanists and scientists but between technology and everybody else. This is not to say that "everybody else" recognizes this. In fact, most people believe that technology is a staunch friend. There are two reasons for this. First, technology *is* a friend. It makes life easier, cleaner, and longer. Can anyone ask more of a friend? Second, because of its lengthy, intimate, and inevitable relationship with culture, technology does not invite a close examination of its own consequences. It is the kind of friend that asks for trust and obedience, which most people are inclined to give because its gifts are truly bountiful. But, of course, there is a dark side to this friend. Its gifts are not without a heavy cost. Stated in the most dramatic terms, the accusation can be made that the uncontrolled growth of technology destroys the vital sources of our humanity. It creates a culture without a moral foundation. It undermines certain mental processes and social relations that make human life worth living. Technology, in sum, is both friend and enemy.

This book attempts to describe when, how, and why technology became a particularly dangerous enemy. The case has been argued many times before by authors of great learning and conviction—in our own time by Lewis Mumford, Jacques Ellul, Herbert Read, Arnold Gehlen, Ivan Illich, to name a few. The argument was interrupted only briefly by Snow's irrelevancies and has continued into our own time with a sense of urgency, made even more compelling by America's spectacular display of technological pre-eminence in the Iraqi war. I do not say here that the war was unjustified or that the technology was misused, only that the American success may serve as a confirmation of the catastrophic idea that in peace as well as war technology will be our savior.

Technopoly

1

The Judgment of Thamus

You will find in Plato's *Phaedrus* a story about Thamus, the king of a great city of Upper Egypt. For people such as ourselves, who are inclined (in Thoreau's phrase) to be tools of our tools, few legends are more instructive than his. The story, as Socrates tells it to his friend Phaedrus, unfolds in the following way: Thamus once entertained the god Theuth, who was the inventor of many things, including number, calculation, geometry, astronomy, and writing. Theuth exhibited his inventions to King Thamus, claiming that they should be made widely known and available to Egyptians. Socrates continues:

> Thamus inquired into the use of each of them, and as Theuth went through them expressed approval or disapproval, according as he judged Theuth's claims to be well or ill founded. It would take too long to go through all that Thamus is reported to have said for and against each of Theuth's inventions. But when it came to writing, Theuth declared, "Here is an accomplishment, my lord the King,

> which will improve both the wisdom and the memory of the Egyptians. I have discovered a sure receipt for memory and wisdom." To this, Thamus replied, "Theuth, my paragon of inventors, the discoverer of an art is not the best judge of the good or harm which will accrue to those who practice it. So it is in this; you, who are the father of writing, have out of fondness for your off-spring attributed to it quite the opposite of its real function. Those who acquire it will cease to exercise their memory and become forgetful; they will rely on writing to bring things to their remembrance by external signs instead of by their own internal resources. What you have discovered is a receipt for recollection, not for memory. And as for wisdom, your pupils will have the reputation for it without the reality: they will receive a quantity of information without proper instruction, and in consequence be thought very knowledgeable when they are for the most part quite ignorant. And because they are filled with the conceit of wisdom instead of real wisdom they will be a burden to society."[1]

I begin my book with this legend because in Thamus' response there are several sound principles from which we may begin to learn how to think with wise circumspection about a technological society. In fact, there is even one error in the judgment of Thamus, from which we may also learn something of importance. The error is not in his claim that writing will damage memory and create false wisdom. It is demonstrable that writing has had such an effect. Thamus' error is in his believing that writing will be a burden to society and *nothing but a burden.* For all his wisdom, he fails to imagine what writing's benefits might be, which, as we know, have been considerable. We may learn from this that it is a mistake to suppose that any technological innovation has a one-sided effect. Every technology is both a burden and a blessing; not either-or, but this-and-that.

Nothing could be more obvious, of course, especially to those who have given more than two minutes of thought to the matter. Nonetheless, we are currently surrounded by throngs of zealous Theuths, one-eyed prophets who see only what new technologies can do and are incapable of imagining what they will *undo.* We might call such people Technophiles. They gaze on technology as a lover does on his beloved, seeing it as without blemish and entertaining no apprehension for the future. They are therefore dangerous and are to be approached cautiously. On the other hand, some one-eyed prophets, such as I (or so I am accused), are inclined to speak only of burdens (in the manner of Thamus) and are silent about the opportunities that new technologies make possible. The Technophiles must speak for themselves, and do so all over the place. My defense is that a dissenting voice is sometimes needed to moderate the din made by the enthusiastic multitudes. If one is to err, it is better to err on the side of Thamusian skepticism. But it is an error nonetheless. And I might note that, with the exception of his judgment on writing, Thamus does not repeat this error. You might notice on rereading the legend that he gives arguments *for* and *against* each of Theuth's inventions. For it is inescapable that every culture must negotiate with technology, whether it does so intelligently or not. A bargain is struck in which technology giveth and technology taketh away. The wise know this well, and are rarely impressed by dramatic technological changes, and never overjoyed. Here, for example, is Freud on the matter, from his doleful *Civilization and Its Discontents:*

> One would like to ask: is there, then, no positive gain in pleasure, no unequivocal increase in my feeling of happiness, if I can, as often as I please, hear the voice of a child

of mine who is living hundreds of miles away or if I can learn in the shortest possible time after a friend has reached his destination that he has come through the long and difficult voyage unharmed? Does it mean nothing that medicine has succeeded in enormously reducing infant mortality and the danger of infection for women in child-birth, and, indeed, in considerably lengthening the average life of a civilized man?

Freud knew full well that technical and scientific advances are not to be taken lightly, which is why he begins this passage by acknowledging them. But he ends it by reminding us of what they have undone:

> If there had been no railway to conquer distances, my child would never have left his native town and I should need no telephone to hear his voice; if travelling across the ocean by ship had not been introduced, my friend would not have embarked on his sea-voyage and I should not need a cable to relieve my anxiety about him. What is the use of reducing infantile mortality when it is precisely that reduction which imposes the greatest restraint on us in the begetting of children, so that, taken all round, we never-theless rear no more children than in the days before the reign of hygiene, while at the same time we have created difficult conditions for our sexual life in marriage. . . . And, finally, what good to us is a long life if it is difficult and barren of joys, and if it is so full of misery that we can only welcome death as a deliverer?[2]

In tabulating the cost of technological progress, Freud takes a rather depressing line, that of a man who agrees with Tho-reau's remark that our inventions are but improved means to an unimproved end. The Technophile would surely answer Freud by saying that life has always been barren of joys and full of misery but that the telephone, ocean liners, and especially the reign of hygiene have not only lengthened life but made it a more agreeable proposition. That is certainly an argument I would make (thus proving I am no one-eyed Technophobe), but it is not necessary at this point to pursue it. I have brought Freud into the conversation only to show that a wise man—even one of such a woeful countenance—must begin his critique of technology by acknowledging its successes. Had King Thamus been as wise as reputed, he would not have forgotten to include in his judgment a prophecy about the powers that writing would enlarge. There is a calculus of technological change that requires a measure of even-handedness.

So much for Thamus' error of omission. There is another omission worthy of note, but it is no error. Thamus simply takes for granted—and therefore does not feel it necessary to say—that writing is not a neutral technology whose good or harm depends on the uses made of it. He knows that the uses made of any technology are largely determined by the structure of the technology itself—that is, that its functions follow from its form. This is why Thamus is concerned not with *what* people will write; he is concerned *that* people will write. It is absurd to imagine Thamus advising, in the manner of today's standard-brand Technophiles, that, if only writing would be used for the production of certain kinds of texts and not others (let us say, for dramatic literature but not for history or philosophy), its disruptions could be minimized. He would regard such counsel as extreme naïveté. He would allow, I imagine, that a technology may be barred entry to a culture. But we may learn from Thamus the following: once a technology is admitted, it plays out its hand; it does what it is designed to do. Our task is to understand what that design is—that is to say, when we admit a new technology to the culture, we must do so with our eyes wide open.

All of this we may infer from Thamus' silence. But we may learn even more from what he does say than from what he doesn't. He points out, for example, that writing will change what is meant by the words "memory" and "wisdom." He fears that memory will be confused with what he disdainfully calls "recollection," and he worries that wisdom will become indistinguishable from mere knowledge. This judgment we must take to heart, for it is a certainty that radical technologies create new definitions of old terms, and that this process takes place without our being fully conscious of it. Thus, it is insidious and dangerous, quite different from the process whereby new technologies introduce new terms to the language. In our own time, we have consciously added to our language thousands of new words and phrases having to do with new technologies—"VCR," "binary digit," "software," "front-wheel drive," "window of opportunity," "Walkman," etc. We are not taken by surprise at this. New things require new words. But new things also modify old words, words that have deep-rooted meanings. The telegraph and the penny press changed what we once meant by "information." Television changes what we once meant by the terms "political debate," "news," and "public opinion." The computer changes "information" once again. Writing changed what we once meant by "truth" and "law"; printing changed them again, and now television and the computer change them once more. Such changes occur quickly, surely, and, in a sense, silently. Lexicographers hold no plebiscites on the matter. No manuals are written to explain what is happening, and the schools are oblivious to it. The old words still look the same, are still used in the same kinds of sentences. But they do not have the same meanings; in some cases, they have opposite meanings. And this is what Thamus wishes to teach us—that technology imperiously commandeers our most important terminology. It redefines "freedom," "truth," "intelligence," "fact," "wisdom," "memory," "history"—all the words we live by. And it does not pause to tell us. And we do not pause to ask.

This fact about technological change requires some elaboration, and I will return to the matter in a later chapter. Here, there are several more principles to be mined from the judgment of Thamus that require mentioning because they presage all I will write about. For instance, Thamus warns that the pupils of Theuth will develop an undeserved reputation for wisdom. He means to say that those who cultivate competence in the use of a new technology become an elite group that are granted undeserved authority and prestige by those who have no such competence. There are different ways of expressing the interesting implications of this fact. Harold Innis, the father of modern communication studies, repeatedly spoke of the "knowledge monopolies" created by important technologies. He meant precisely what Thamus had in mind: those who have control over the workings of a particular technology accumulate power and inevitably form a kind of conspiracy against those who have no access to the specialized knowledge made available by the technology. In his book *The Bias of Communication,* Innis provides many historical examples of how a new technology "busted up" a traditional knowledge monopoly and created a new one presided over by a different group. Another way of saying this is that the benefits and deficits of a new technology are not distributed equally. There are, as it were, winners and losers. It is both puzzling and poignant that on many occasions the losers, out of ignorance, have actually cheered the winners, and some still do.

Let us take as an example the case of television. In the United States, where television has taken hold more deeply than anywhere else, many people find it a blessing, not least those who have achieved high-paying, gratifying careers in television as executives, technicians, newscasters, and entertainers. It should surprise no one that such people, forming as they do a new knowledge monopoly, should cheer themselves and defend and promote television technology. On the other hand and in the long run, television may bring a gradual end to the careers of schoolteachers, since school was an invention of the printing press and must stand or fall on the issue of how much importance the printed word has. For four hundred years, schoolteachers have been part of the knowledge monopoly created by printing, and they are now witnessing the breakup of that monopoly. It appears as if they can do little to prevent that breakup, but surely there is something perverse about schoolteachers' being enthusiastic about what is happening. Such enthusiasm always calls to my mind an image of some turn-of-the-century blacksmith who not only sings the praises of the automobile but also believes that his business will be enhanced by it. We know now that his business was not enhanced by it; it was rendered obsolete by it, as perhaps the clearheaded blacksmiths knew. What could they have done? Weep, if nothing else.

We have a similar situation in the development and spread of computer technology, for here too there are winners and losers. There can be no disputing that the computer has increased the power of large-scale organizations like the armed forces, or airline companies or banks or tax-collecting agencies. And it is equally clear that the computer is now indispensable to high-level researchers in physics and other natural sciences. But to what extent has computer technology been an advantage to the masses of people? To steelworkers, vegetable-store owners, teachers, garage mechanics, musicians, bricklayers, dentists, and most of the rest into whose lives the computer now intrudes? Their private matters have been made more accessible to powerful institutions. They are more easily tracked and controlled; are subjected to more examinations; are increasingly mystified by the decisions made about them; are often reduced to mere numerical objects. They are inundated by junk mail. They are easy targets for advertising agencies and political organizations. The schools teach their children to operate computerized systems instead of teaching things that are more valuable to children. In a word, almost nothing that they need happens to the losers. Which is why they are losers.

It is to be expected that the winners will encourage the losers to be enthusiastic about computer technology. That is the way of winners, and so they sometimes tell the losers that with personal computers the average person can balance a checkbook more neatly, keep better track of recipes, and make more logical shopping lists. They also tell them that their lives will be conducted more efficiently. But discreetly they neglect to say from whose point of view the efficiency is warranted or what might be its costs. Should the losers grow skeptical, the winners dazzle them with the wondrous feats of computers, almost all of which have only marginal relevance to the quality of the losers' lives but which are nonetheless impressive. Eventually, the losers succumb, in part because they believe, as Thamus prophesied, that the specialized knowledge of the masters of a new technology is a form of wisdom. The masters come to believe this as well, as Thamus also prophesied. The result is that certain questions do not arise. For example, to whom will the technology give greater power and freedom? And whose power and freedom will be reduced by it?

I have perhaps made all of this sound like a well-planned conspiracy, as if the winners know all too well what is being won and what lost. But this is not quite how it happens. For one thing, in cultures that have a democratic ethos, relatively weak traditions, and a high receptivity to new technologies, everyone is inclined to be enthusiastic about technological change, believing that its benefits will eventually spread evenly among the entire population. Especially in the United States, where the lust for what is new has no bounds, do we find this childlike conviction most widely held. Indeed, in America, social change of any kind is rarely seen as resulting in winners and losers, a condition that stems in part from Americans' much-documented optimism. As for change brought on by technology, this native optimism is exploited by entrepreneurs, who work hard to infuse the population with a unity of improbable hope, for they know that it is economically unwise to reveal the price to be paid for technological change. One might say, then, that, if there is a conspiracy of any kind, it is that of a culture conspiring against itself.

In addition to this, and more important, it is not always clear, at least in the early stages of a technology's intrusion into a culture, who will gain most by it and who will lose most. This is because the changes wrought by technology are subtle if not downright mysterious, one might even say wildly unpredictable. Among the most unpredictable are those that might be labeled ideological. This is the sort of change Thamus had in mind when he warned that writers will come to rely on external signs instead of their own internal resources, and that they will receive quantities of information without proper instruction. He meant that new technologies change what we mean by "knowing" and "truth"; they alter those deeply embedded habits of thought which give to a culture its sense of what the world is like—a sense of what is the natural order of things, of what is reasonable, of what is necessary, of what is inevitable, of what is real. Since such changes are expressed in changed meanings of old words, I will hold off until later discussing the massive ideological transformation now occurring in the United States. Here, I should like to give only one example of how technology creates new conceptions of what is real and, in the process, undermines older conceptions. I refer to the seemingly harmless practice of assigning marks or grades to the answers students give on examinations. This procedure seems so natural to most of us that we are hardly aware of its significance. We may even find it difficult to imagine that the number or letter is a tool or, if you will, a technology; still less that, when we use such a technology to judge someone's behavior, we have done something peculiar. In point of fact, the first instance of grading students' papers occurred at Cambridge University in 1792 at the suggestion of a tutor named William Farish.[3] No one knows much about William Farish; not more than a handful have ever heard of him. And yet his idea that a quantitative value should be assigned to human thoughts was a major step toward constructing a mathematical concept of reality. If a number can be given to the quality of a thought, then a number can be given to the qualities of mercy, love, hate, beauty, creativity, intelligence, even sanity itself. When Galileo said that the language of nature is written in mathematics, he did not mean to include human feeling or accomplishment or insight. But most of us are now inclined to make these inclusions. Our psychologists, sociologists, and educators find it quite impossible to do their work without numbers. They believe that without numbers they cannot acquire or express authentic knowledge.

I shall not argue here that this is a stupid or dangerous idea, only that it is peculiar. What is even more peculiar is that so many of us do not find the idea peculiar. To say that someone should be doing better work because he has an IQ of 134, or that someone is a 7.2 on a sensitivity scale, or that this man's essay on the rise of capitalism is an A− and that man's is a C+ would have sounded like gibberish to Galileo or Shakespeare or Thomas Jefferson. If it makes sense to us, that is because our minds have been conditioned by the technology of numbers so that we see the world differently than they did. Our understanding of what is real is different. Which is another way of saying that embedded in every tool is an ideological bias, a predisposition to construct the world as one thing rather than another, to value one thing over another, to amplify one sense or skill or attitude more loudly than another.

This is what Marshall McLuhan meant by his famous aphorism "The medium is the message." This is what Marx meant when he said, "Technology discloses man's mode of dealing with nature" and creates the "conditions of intercourse" by which we relate to each other. It is what Wittgenstein meant when, in referring to our most fundamental technology, he said that language is not merely a vehicle of thought but also the driver. And it is what Thamus wished the inventor Theuth to see. This is, in short, an ancient and persistent piece of wisdom, perhaps most simply expressed in the old adage that, to a man with a hammer, everything looks like a nail. Without being too literal, we may extend the truism: To a man with a pencil, everything looks like a list. To a man with a camera, everything looks like an image. To a man with a computer, everything looks like data. And to a man with a grade sheet, everything looks like a number.

But such prejudices are not always apparent at the start of a technology's journey, which is why no one can safely conspire to be a winner in technological change. Who would have imagined, for example, whose interests and what world-view would be ultimately advanced by the invention of the mechanical clock? The clock had its origin in the Benedictine monasteries of the twelfth and thirteenth centuries. The impetus behind the invention was to provide a more or less precise regularity to the routines of the monasteries, which required, among other things, seven periods of devotion during the course of the day. The bells of the monastery were to be rung to signal the canonical hours; the mechanical clock was the technology that could provide precision to these rituals of devotion. And indeed it did. But what the monks did not foresee was that the clock is a means not merely of keeping track of the hours but also of synchronizing and controlling the actions of men. And thus, by the middle of the fourteenth century, the clock had moved outside the walls of the monastery, and brought a new and precise regularity to the life of the workman and the merchant.

"The mechanical clock," as Lewis Mumford wrote, "made possible the idea of regular production, regular working hours and a standardized product." In short, without the clock, capitalism would have been quite impossible.[4] The paradox, the surprise, and the wonder are that the clock was invented by men who wanted to devote themselves more rigorously to God; it ended as the technology of greatest use to men who wished to devote themselves to the accumulation of money. In the eternal struggle between God and Mammon, the clock quite unpredictably favored the latter.

Unforeseen consequences stand in the way of all those who think they see clearly the direction in which a new technology will take us. Not even those who invent a technology can be assumed to be reliable prophets, as Thamus warned. Gutenberg, for example, was by all accounts a devout Catholic who would have been horrified to hear that accursed heretic Luther describe printing as "God's highest act of grace, whereby the business of the Gospel is driven forward." Luther understood, as Gutenberg did not, that the mass-produced book, by placing the Word of God on every kitchen table, makes each Christian his own theologian—one might even say his own priest, or, better, from Luther's point of view, his own pope. In the struggle between unity and diversity of religious belief, the press favored the latter, and we can assume that this possibility never occurred to Gutenberg.

Thamus understood well the limitations of inventors in grasping the social and psychological—that is, ideological—bias of their own inventions. We can imagine him addressing Gutenberg in the following way: "Gutenberg, my paragon of inventors, the discoverer of an art is not the best judge of the good or harm which will accrue to those who practice it. So it is in this; you, who are the father of printing, have out of fondness for your off-spring come to believe it will advance the cause of the Holy Roman See, whereas in fact it will sow discord among believers; it will damage the authenticity of your beloved Church and destroy its monopoly."

We can imagine that Thamus would also have pointed out to Gutenberg, as he did to Theuth, that the new invention would create a vast population of readers who "will receive a quantity of information without proper instruction . . . [who will be] filled with the conceit of wisdom instead of real wisdom"; that reading, in other words, will compete with older forms of learning. This is yet another principle of technological change we may infer from the judgment of Thamus: new technologies compete with old ones—for time, for attention, for money, for prestige, but mostly for dominance of their world-view. This competition is implicit once we acknowledge that a medium contains an ideological bias. And it is a fierce competition, as only ideological competitions can be. It is not merely a matter of tool against tool—the alphabet attacking ideographic writing, the printing press attacking the illuminated manuscript, the photograph attacking the art of painting, television attacking the printed word. When media make war against each other, it is a case of world-views in collision.

In the United States, we can see such collisions everywhere—in politics, in religion, in commerce—but we see them most clearly in the schools, where two great technologies confront each other in uncompromising aspect for the control of students' minds. On the one hand, there is the world of the printed word with its emphasis on logic, sequence, history, exposition, objectivity, detachment, and discipline. On the other, there is the world of television with its emphasis on imagery, narrative, presentness, simultaneity, intimacy, immediate gratification, and quick emotional response. Children come to school having been deeply conditioned by the biases of television. There, they encounter the world of the printed word. A sort of psychic battle takes place, and there are many casualties—children who can't learn to read or won't, children who cannot organize their thought into logical structure even in a simple paragraph, children who cannot attend to lectures or oral explanations for more than a few minutes at a time. They are failures, but not because they are stupid. They are failures because there is a media war going on, and they are on the wrong side—at least for the moment. Who knows what schools will be like twenty-five years from now? Or fifty? In time, the type of student who is currently a failure may be considered a success. The type who is now successful may be regarded as a handicapped learner—slow to respond, far too detached, lacking in emotion, inadequate in creating mental pictures of reality. Consider: what Thamus called the "conceit of wisdom"—the unreal knowledge acquired through the written word—eventually became the pre-eminent form of knowledge valued by the schools. There is no reason to suppose that such a form of knowledge must always remain so highly valued.

To take another example: In introducing the personal computer to the classroom, we shall be breaking a four-hundred-year-old truce between the gregariousness and openness fostered by orality and the introspection and isolation fostered by the printed word. Orality stresses group learning, cooperation, and a sense of social responsibility, which is the context within which Thamus believed proper instruction and real knowledge must be communicated. Print stresses individualized learning, competition, and personal autonomy. Over four centuries, teachers, while emphasizing print, have allowed orality its place in the classroom, and have therefore achieved a kind of pedagogical peace between these two forms of learning, so that what is valuable in each can be maximized. Now comes the computer, carrying anew the banner of private learning and individual problem-solving. Will the widespread use of computers in the classroom defeat once and for all the claims of communal speech? Will the computer raise egocentrism to the status of a virtue?

These are the kinds of questions that technological change brings to mind when one grasps, as Thamus did, that technological competition ignites total war, which means it is not possible to contain the effects of a new technology to a limited sphere of human activity. If this metaphor puts the matter too brutally, we may try a gentler, kinder one: Technological change is neither additive nor subtractive. It is ecological. I mean "ecological" in the same sense as the word is used by environmental scientists. One significant change generates total change. If you remove the caterpillars from a given habitat, you are not left with the same environment minus caterpillars: you have a new environment, and you have reconstituted the conditions of survival; the same is true if you add caterpillars to an environment that has had none. This is how the ecology of media works as well. A new technology does not add or subtract something. It changes everything. In the year 1500, fifty years after the printing press was invented, we did not have old Europe plus the printing press. We had a different Europe. After television, the United States was not America plus television; television gave a new coloration to every political campaign, to every home, to every school, to every church, to every industry. And that is why the competition among media is so fierce. Surrounding every technology are institutions whose organization—not to mention their reason for being—reflects the world-view promoted by the technology. Therefore, when an old technology is assaulted by a new one, institutions are threatened. When institutions are threatened, a culture finds itself in crisis. This is serious business, which is why we learn nothing when educators ask, Will students learn mathematics better by computers than by textbooks? Or when businessmen ask, Through which medium can we sell more products? Or when preachers ask, Can we reach more people through television than through radio? Or when politicians ask, How effective are messages sent through different media? Such questions have an immediate, practical value to those who ask them, but they are diversionary. They direct our attention away from the serious social, intellectual, and institutional crises that new media foster.

Perhaps an analogy here will help to underline the point. In speaking of the meaning of a poem, T. S. Eliot remarked that the chief use of the overt content of poetry is "to satisfy one habit of the reader, to keep his mind diverted and quiet, while the poem does its work upon him: much as the imaginary burglar is always provided with a bit of nice meat for the house-dog." In other words, in asking their practical questions, educators, entrepreneurs, preachers, and politicians are like the house-dog munching peacefully on the meat while the house is looted. Perhaps some of them know this and do not especially care. After all, a nice piece of meat, offered graciously, does take care of the problem of where the next meal will come from. But for the rest of us, it cannot be acceptable to have the house invaded without protest or at least awareness.

What we need to consider about the computer has nothing to do with its efficiency as a teaching tool. We need to know in what ways it is altering our conception of learning, and how, in conjunction with television, it undermines the old idea of school. Who cares how many boxes of cereal can be sold via television? We need to know if television changes our conception of reality, the relationship of the rich to the poor, the idea of happiness itself. A preacher who confines himself to considering how a medium can increase his audience will miss the significant question: In what sense do new media alter what is meant by religion, by church, even by God? And if the politician cannot think beyond the next election, then *we* must wonder about what new media do to the idea of political organization and to the conception of citizenship.

To help us do this, we have the judgment of Thamus, who, in the way of legends, teaches us what Harold Innis, in his way, tried to. New technologies alter the structure of our interests: the things we think *about.* They alter the character of our symbols: the things we think *with.* And they alter the nature of community: the arena in which thoughts develop. As Thamus spoke to Innis across the centuries, it is essential that we listen to their conversation, join in it, revitalize it. For something has happened in America that is strange and dangerous, and there is only a dull and even stupid awareness of what it is—in part because it has no name. I call it Technopoly.

2

From Tools to Technocracy

Among the famous aphorisms from the troublesome pen of Karl Marx is his remark in *The Poverty of Philosophy* that the "hand-loom gives you society with the feudal lord; the steam-mill, society with the industrial capitalist." As far as I know, Marx did not say which technology gives us the technocrat, and I am certain his vision did not include the emergence of the Technopolist. Nonetheless, the remark is useful. Marx understood well that, apart from their economic implications, technologies create the ways in which people perceive reality, and that such ways are the key to understanding diverse forms of social and mental life. In *The German Ideology,* he says, "As individuals express their life, so they are," which sounds as much like Marshall McLuhan or, for that matter, Thamus as it is possible to sound. Indeed, toward the end of that book, Marx includes a remarkable paragraph that would be entirely at home in McLuhan's *Understanding Media.* "Is Achilles possible," he asks, "when powder and shot have been invented? And is the Iliad possible at all when the printing press and even printing machines exist? Is it not inevitable that with the emergence of the press, the singing and the telling and the muse cease; that is, the conditions for epic poetry disappear?"[1]

By connecting technological conditions to symbolic life and psychic habits, Marx was doing nothing unusual. Before him, scholars found it useful to invent taxonomies of culture based on the technological character of an age. And they do it still, for the practice is something of a persistent scholarly industry. We think at once of the best-known classification: the Stone Age, the Bronze Age, the Iron Age, the Steel Age. We speak easily of the Industrial Revolution, a term popularized by Arnold Toynbee, and, more recently, of the Post-Industrial Revolution, so named by Daniel Bell. Oswald Spengler wrote of the Age of Machine Technics, and C. S. Peirce called the nineteenth century the Railway Age. Lewis Mumford, looking at matters from a longer perspective, gave us the Eotechnic, the Paleotechnic, and the Neotechnic Ages. With equally telescopic perspective, José Ortega y Gasset wrote of three stages in the development of technology: the age of technology of chance, the age of technology of the artisan, the age of technology of the technician. Walter Ong has written about Oral cultures, Chirographic cultures, Typographic cultures, and Electronic cultures. McLuhan himself introduced the phrase "the Age of Gutenberg" (which, he believed, is now replaced by the Age of Electronic Communication).

I find it necessary, for the purpose of clarifying our present situation and indicating what dangers lie ahead, to create still another taxonomy. Cultures may be classified into three types: tool-using cultures, technocracies, and technopolies. At the present time, each type may be found somewhere on the planet, although the first is rapidly disappearing: we must travel to exotic places to find a tool-using culture.[2] If we do, it is well to go armed with the knowledge that, until the seventeenth century, all cultures were tool-users. There was, of course, considerable variation from one culture to another in the tools that were available. Some had only spears and cooking utensils. Some had water mills and coal- and horsepower. But the main characteristic of all tool-using cultures is that their tools were largely invented to do two things: to solve specific and urgent problems of physical life, such as in the use of waterpower, windmills, and the heavy-wheeled plow; or to serve the symbolic world of art, politics, myth, ritual, and religion, as in the construction of castles and cathedrals and the development of the mechanical clock. In either case, tools did not attack (or, more precisely, were not intended to attack) the dignity and integrity of the culture into which they were introduced. With some exceptions, tools did not prevent people from believing in their traditions, in their God, in their politics, in their methods of education, or in the legitimacy of their social organization. These beliefs, in fact, *directed* the invention of tools and limited the uses to which they were put. Even in the case of military technology, spiritual ideas and social customs acted as controlling forces. It is well known, for example, that the uses of the sword by samurai warriors were meticulously governed by a set of ideals known as Bushido, or the Way of the Warrior. The rules and rituals specifying when, where, and how the warrior must use either of his two swords (the *katana,* or long sword, and the *wakizashi,* or short sword) were precise, tied closely to the concept of honor, and included the requirement that the warrior commit seppuku or hara-kiri should his honor be compromised. This sort of governance of military technology was not unknown in the Western world. The use of the lethal crossbow was prohibited, under threat of anathema, by Pope Innocent II in the early twelfth century. The weapon was judged to be "hateful to God" and therefore could not be used against Christians. That it could be used against Muslims and other infidels does not invalidate the point that in a tool-using culture technology is not seen as autonomous, and is subject to the jurisdiction of some binding social or religious system.

Having defined tool-using cultures in this manner, I must add two points so as to avoid excessive oversimplification. First, the quantity of technologies available to a tool-using culture is not its defining characteristic. Even a superficial study of the Roman Empire, for example, reveals the extent to which it relied on roads, bridges, aqueducts, tunnels, and sewers for both its economic vitality and its military conquests. Or, to take another example, we know that, between the tenth and thirteenth centuries, Europe underwent a technological boom: medieval man was surrounded by machines.[3] One may even go as far as Lynn White, Jr., who said that the Middle Ages gave us for the first time in history "a complex civilization which rested not on the backs of sweating slaves or coolies but primarily on non-human power."[4] Tool-using cultures, in other words, may be both ingenious and productive in solving problems of the physical environment. Windmills were invented in the late twelfth century. Eyeglasses for nearsightedness appeared in Italy in 1280. The invention in the eleventh century of rigid padded collars to rest on the shoulder blades of horses solved the problem of how to increase the pulling power of horses without decreasing their ability to breathe. In fact, as early as the ninth century in Europe, horseshoes were invented, and someone figured out that, when horses are hitched, one behind the other, their pulling power is enormously amplified. Corn mills, paper mills, and fulling mills were part of medieval culture, as were bridges, castles, and cathedrals. The famous spire of Strasbourg Cathedral, built in the thirteenth century, rose to a height of 466 feet, the equivalent of a forty-story skyscraper. And, to go further back in time, one must not fail to mention the remarkable engineering achievements of Stonehenge and the Pyramids (whose construction, Lewis Mumford insisted, signifies the first example of a megamachine in action).

Given the facts, we must conclude that tool-using cultures are not necessarily impoverished technologically, and may even be surprisingly sophisticated. Of course, some tool-using cultures were (and still are) technologically primitive, and some have even displayed a contempt for crafts and machinery. The Golden Age of Greece, for example, produced no important technical inventions and could not even devise ways of using horsepower efficiently. Both Plato and Aristotle scorned the "base mechanic arts," probably in the belief that nobility of mind was not enhanced by efforts to increase efficiency or productivity. Efficiency and productivity were problems for slaves, not philosophers. We find a somewhat similar view in the Bible, which is the longest and most detailed account of an ancient tool-using culture we have. In Deuteronomy, no less an authority than God Himself says, "Cursed be the man who makes a graven or molten image, an abomination to the Lord, a thing made by the hands of a craftsman, and sets it up in secret."

Tool-using cultures, then, may have many tools or few, may be enthusiastic about tools or contemptuous. The name "tool-using culture" derives from the relationship in a given culture between tools and the belief system or ideology. The tools are not intruders. They are integrated into the culture in ways that do not pose significant contradictions to its world-view. If we take the European Middle Ages as an example of a tool-using culture, we find a very high degree of integration between its tools and its world-view. Medieval theologians developed an elaborate and systematic description of the relation of man to God, man to nature, man to man, and man to his tools. Their theology took as a first and last principle that all knowledge and goodness come from God, and that therefore all human enterprise must be directed toward the service of God. Theology, not technology, provided people with authorization for what to do or think. Perhaps this is why Leonardo da Vinci kept his design of a submarine secret, believing that it was too harmful a tool to unleash, that it would not gain favor in God's eyes.

In any case, theological assumptions served as a controlling ideology, and whatever tools were invented had, ultimately, to fit within that ideology. We may say, further, that all tool-using cultures—from the technologically most primitive to the most sophisticated—are theocratic or, if not that, unified by some metaphysical theory. Such a theology or metaphysics provides order and meaning to existence, making it almost impossible for technics to subordinate people to its own needs.

The "almost" is important. It leads to my second qualification. As the spirit of Thamus reminds us, tools have a way of intruding on even the most unified set of cultural beliefs. There are limits to the power of both theology and metaphysics, and technology has business to do which sometimes cannot be stayed by any force. Perhaps the most interesting example of a drastic technological disruption of a tool-using culture is in the eighth-century use of the stirrup by the Franks under the leadership of Charles Martel. Until this time, the principal use of horses in combat was to transport warriors to the scene of the battle, whereupon they dismounted to meet the foe. The stirrup made it possible to fight *on* horseback, and this created an awesome new military technology: mounted shock combat. The new form of combat, as Lynn White, Jr., has meticulously detailed, enlarged the importance of the knightly class and changed the nature of feudal society.[5] Landholders found it necessary to secure the services of cavalry for protection. Eventually, the knights seized control of church lands and distributed them to vassals on condition that they stay in the service of the knights. If a pun will be allowed here, the stirrup was in the saddle, and took feudal society where it would not otherwise have gone.

To take a later example: I have already alluded to the transformation of the mechanical clock in the fourteenth century from an instrument of religious observance to an instrument of commercial enterprise. That transformation is sometimes given a specific date—1370—when King Charles V ordered all citizens of Paris to regulate their private, commercial, and industrial life by the bells of the Royal Palace clock, which struck every sixty minutes. All churches in Paris were similarly required to regulate their clocks, in disregard of the canonical hours. Thus, the church had to give material interests precedence over spiritual needs. Here is a clear example of a tool being employed to loosen the authority of the central institution of medieval life.

There are other examples of how technologies created problems for the spiritual life of medieval Europe. For example, the mills to which farmers flocked to have their grain ground became a favorite place for prostitutes to attract customers. The problem grew to such proportions that Saint Bernard, the leader of the Cistercian order in the twelfth century, tried to close down the mills. He was unsuccessful, because the mills had become too important to the economy. In other words, it is something of an oversimplification to say that tool-using cultures never had their customs and symbolic life reoriented by technology. And, just as there are examples of such cases in the medieval world, we can find queer but significant instances in technologically primitive societies of tools attacking the supremacy of custom, religion, or metaphysics. Egbert de Vries, a Dutch sociologist, has told of how the introduction of matches to an African tribe altered their sexual habits.[6] Members of this community believed it necessary to start a new fire in the fireplace after each act of sexual intercourse. This custom meant that each act of intercourse was something of a public event, since when it was completed someone had to go to a neighboring hut to bring back a burning stick with which to start a fresh fire. Under such conditions, adultery was difficult to conceal, which is conceivably why the custom originated in the first place. The introduction of matches changed all this. It became possible to light a new fire without going to a neighbor's hut, and thus, in a flash, so to speak, a long-standing tradition was consumed. In reporting on de Vries' finding, Alvin Toffler raises several intriguing questions: Did matches result in a shift in values? Was adultery less or more frowned upon as a result? By facilitating the privacy of sex, did matches alter the valuation placed upon it? We can be sure that some changes in cultural values occurred, although they could not have been as drastic as what happened to the Ihalmiut tribe early in the twentieth century, after the introduction of the rifle. As described by Farley Mowat in *The People of the Deer*, the replacement of bows and arrows with rifles is one of the most chilling tales on record of a technological attack on a tool-using culture. The result in this case was not the modification of a culture but its eradication.

Nonetheless, after one acknowledges that no taxonomy ever neatly fits the realities of a situation, and that in particular the definition of a tool-using culture lacks precision, it is still both possible and useful to distinguish a tool-using culture from a technocracy. In a technocracy, tools play a central role in the thought-world of the culture. Everything must give way, in some degree, to their development. The social and symbolic worlds become increasingly subject to the requirements of that development. Tools are not integrated into the culture; they attack the culture. They bid to *become* the culture. As a consequence, tradition, social mores, myth, politics, ritual, and religion have to fight for their lives.

The modern technocracies of the West have their roots in the medieval European world, from which there emerged three great inventions: the mechanical clock, which provided a new conception of time; the printing press with movable type, which attacked the epistemology of the oral tradition; and the telescope, which attacked the fundamental propositions of Judeo-Christian theology. Each of these was significant in creating a new relationship between tools and culture. But since it is permissible to say that among faith, hope, and charity the last is most important, I shall venture to say that among the clock, the press, and the telescope the last is also the most important. To be more exact (since Copernicus, Tycho Brahe, and to some extent Kepler did their work without benefit of the telescope), somewhat cruder instruments of observation than the telescope allowed men to see, measure, and speculate about the heavens in ways that had not been possible before. But the refinements of the telescope made their knowledge so precise that there followed a collapse, if one may say it this way, of the moral center of gravity in the West. That moral center had allowed people to believe that the earth was the stable center of the universe and therefore that humankind was of special interest to God. After Copernicus, Kepler, and especially Galileo, the Earth became a lonely wanderer in an obscure galaxy in some hidden corner of the universe, and this left the Western world to wonder if God had any interest in us at all. Although John Milton was only an infant when Galileo's *Messenger from the Stars* was printed in 1610, he was able, years later, to describe the psychic desolation of an unfathomable universe that Galileo's telescopic vision thrust upon an unprepared theology. In *Paradise Lost*, Milton wrote:

> *Before [his] eyes in sudden view appear*
> *The secrets of the hoary Deep—a dark*
> *Illimitable ocean, without bound,*
> *Without dimension. . . .*

Truly, a paradise lost. But it was not Galileo's intention—neither was it Copernicus' or Kepler's—to so disarm their culture. These were medieval men who, like Gutenberg before them, had no wish to damage the spiritual foundations of their world. Copernicus, for example, was a doctor of canon law, having been elected a canon of Frauenburg Cathedral. Although he never took a medical degree, he studied medicine, was private physician to his uncle, and among many people was better known as a physician than as an astronomer. He published only one scientific work, *On the Revolutions of the Heavenly Spheres,* the first completed copy arriving from the printer only a few hours before his death, at the age of seventy, on May 24, 1543. He had delayed publishing his heliocentric theory for thirty years, largely because he believed it to be unsound, not because he feared retribution from the church. In fact, his book was not placed on the Index until seventy-three years after it was published, and then only for a short time. (Galileo's trial did not take place until ninety years after Copernicus' death.) In 1543, scholars and philosophers had no reason to fear persecution for their ideas so long as they did not directly challenge the authority of the church, which Copernicus had no wish to do. Though the authorship of the preface to his work is in dispute, the preface clearly indicates that his ideas are to be taken as hypotheses, and that his "hypotheses need not be true or even probable." We can be sure that Copernicus believed that the earth really moved, but he did not believe that either the earth or the planets moved in the manner described in his system, which he understood to consist of geometric fictions. And he did not believe that his work undermined the supremacy of theology. It is true that Martin Luther called Copernicus "a fool who went against Holy Writ," but Copernicus did not think he had done so—which proves, I suppose, that Luther saw more deeply than Copernicus.

Kepler's is a somewhat similar story. Born in 1571, he began his career by publishing astrological calendars, and ended it as court astrologer to the duke of Wallenstein. Although he was famous for his service as an astrologer, we must credit him with believing that "Astrology can do enormous harm to a monarch if a clever astrologer exploits his human credulity." Kepler wished astrology to be kept out of sight of all heads of state, a precaution that in recent years has not always been taken. His mother was accused of being a witch, and although Kepler did not believe this specific charge, he would probably not have denied categorically the existence of witches. He spent a great deal of his time corresponding with scholars on questions concerning chronology in the age of Christ, and his theory that Jesus was actually born in 4 or 5 B.C. is generally accepted today. In other words, Kepler was very much a man of his time, medieval through and through. Except for one thing: He believed that theology and science should be kept separate and, in particular, that angels, spirits and the opinions of saints should be banished from cosmology. In his *New Astronomy,* he wrote, "Now as regards the opinions of the saints about these matters of nature, I answer in one word, that in theology the weight of authority, but in philosophy the weight of Reason alone is valid." After reviewing what various saints had said about the earth, Kepler concluded, ". . . but to me more sacred than all these is Truth, when I, with all respect for the doctors of the Church, demonstrate from philosophy that the earth is round, circumhabited by antipodes, of a most insignificant smallness, and a swift wanderer among the stars."

In expressing this idea, Kepler was taking the first significant step toward the conception of a technocracy. We have here a clear call for a separation of moral and intellectual values, a separation that is one of the pillars of a technocracy—a significant step but still a small one. No one before Kepler had asked why planets travel at variable rates. Kepler's answer was that it must be a force emanating from the sun. But this answer still had room in it for God. In a famous letter sent to his colleague Maestlin, Kepler wrote, "The sun in the middle of the moving stars, himself at rest and yet the source of motion, carries the image of God the Father and Creator. . . . He distributes his motive force through a medium which contains the moving bodies even as the Father creates through the Holy Ghost."

Kepler was a Lutheran, and although he was eventually excommunicated from his own church, he remained a man of sincere religious conviction to the end. He was, for example, dissatisfied with his discovery of the elliptical orbits of planets, believing that an ellipse had nothing to recommend it in the eyes of God. To be sure, Kepler, building on the work of Copernicus, was creating something new in which truth was not required to gain favor in God's eyes. But it was not altogether clear to him exactly what his work would lead to. It remained for Galileo to make visible the unresolvable contradictions between science and theology, that is, between intellectual and moral points of view.

Galileo did not invent the telescope, although he did not always object to the attribution. A Dutch spectacle-maker named Johann Lippershey was probably the instrument's true inventor; at any rate, he was the first to claim a license for its manufacture, in 1608. (It might also be worth remarking here that the famous experiment of dropping cannon balls from the Tower of Pisa was not only *not* done by Galileo but actually carried out by one of his adversaries, Giorgio Coressio, who was trying to confirm, not dispute, Aristotle's opinion that larger bodies fall more quickly than smaller ones.) Nonetheless, to Galileo must go the entire credit for transforming the telescope from a toy into an instrument of science. And to Galileo must also go the credit of making astronomy a source of pain and confusion to the prevailing theology. His discovery of the four moons of Jupiter and the simplicity and accessibility of his writing style were key weapons in his arsenal. But more important was the directness with which he disputed the Scriptures. In his famous *Letter to the Grand Duchess Christina,* he used arguments first advanced by Kepler as to why the Bible could not be interpreted literally. But he went further in saying that nothing physical that could be directly observed or which demonstrations could prove ought to be questioned merely because Biblical passages say otherwise. More clearly than Kepler had been able to do, Galileo disqualified the doctors of the church from offering opinions about nature. To allow them to do so, he charged, is pure folly. He wrote, "This would be as if an absolute despot, being neither a physician nor an architect, but knowing himself free to command, should undertake to administer medicines and erect buildings according to his whim—at grave peril of his poor patients' lives, and the speedy collapse of his edifices."

From this and other audacious arguments, the doctors of the church were sent reeling. It is therefore astonishing that the church made persistent efforts to accommodate its beliefs to Galileo's observations and claims. It was willing, for example, to accept as hypotheses that the earth moves and that the sun stands still. This, on the grounds that it is the business of mathematicians to formulate interesting hypotheses. But there could be no accommodation with Galileo's claim that the movement of the earth is a fact of nature. Such a belief was definitively held to be injurious to holy faith by contradicting Scripture. Thus, the trial of Galileo for heresy was inevitable even though long delayed. The trial took place in 1633, resulting in Galileo's conviction. Among the punishments were that Galileo was to abjure Copernican opinion, serve time in a formal prison, and for three years repeat once a week seven penitential psalms. There is probably no truth to the belief that Galileo mumbled at the conclusion of his sentencing, "But the earth moves" or some similar expression of defiance. He had, in fact, been asked four times at his trial if he believed in the Copernican view, and each time he said he did not. Everyone knew he believed otherwise, and that it was his advanced age, infirmities, and fear of torture that dictated his compliance. In any case, Galileo did not spend a single day in prison. He was confined at first to the grand duke's villa at Trinità del Monte, then to the palace of Archbishop Piccolomini in Siena, and finally to his home in Florence, where he remained for the rest of his life. He died in 1642, the year Isaac Newton was born.

Copernicus, Kepler, and Galileo put in place the dynamite that would blow up the theology and metaphysics of the medieval world. Newton lit the fuse. In the ensuing explosion, Aristotle's animism was destroyed, along with almost everything else in his *Physics.* Scripture lost much of its authority. Theology, once the Queen of the Sciences, was now reduced to the status of Court Jester. Worst of all, the meaning of existence itself became an open question. And how ironic it all was! Whereas men had traditionally looked to Heaven to find authority, purpose, and meaning, the Sleepwalkers (as Arthur Koestler called Copernicus, Kepler, and Galileo) looked not to Heaven but to the heavens. There they found only mathematical equations and geometric patterns. They did so with courage but not without misgivings, for they did their best to keep their faith, and they did not turn away from God. They believed in a God who had planned and designed the whole of creation, a God who was a master mathematician. Their search for the mathematical laws of nature was, fundamentally, a religious quest. Nature was God's text, and Galileo found that God's alphabet consisted of "triangles, quadrangles, circles, spheres, cones, pyramids, and other mathematical figures." Kepler agreed, and even boasted that God, the author, had to wait six thousand years for His first reader—Kepler himself. As for Newton, he spent most of his later years trying to compute the generations since Adam, his faith in Scripture being unshaken. Descartes, whose *Discourse on Method,* published in 1637, provided nobility to skepticism and reason and served as a foundation of the new science, was a profoundly religious man. Although he saw the universe as mechanistic ("Give me matter and motion," he wrote, "and I will construct the world"), he deduced his law of the immutability of motion from the immutability of God.

All of them, to the end, clung to the theology of their age. They would surely not have been indifferent to knowing when the Last Judgment would come, and they could not have imagined the world without God. Moreover, the science they created was almost wholly concerned with questions of truth, not power. Toward that end, there developed in the late sixteenth century what can only be described as a passion for exactitude: exact dates, quantities, distances, rates. It was even thought possible to determine the exact moment of the Creation, which, as it turned out, commenced at 9:00 a.m., October 23, 4004 B.C. These were men who thought of philosophy (which is what they called science) as the Greeks did, believing that the true object of investigating nature is speculative satisfaction. They were not concerned with the idea of progress, and did not believe that their speculations held the promise of any important improvements in the conditions of life. Copernicus, Kepler, Galileo, Descartes, and Newton laid the foundation for the emergence of technocracies, but they themselves were men of tool-using cultures.

Francis Bacon, born in 1561, was the first man of the technocratic age. In saying this, I may be disputing no less an authority than Immanuel Kant, who said that a Kepler or a Newton was needed to find the law of the movement of civilization. Perhaps. But it was Bacon who first saw, pure and serene, the connection between science and the improvement of the human condition. The principal aim of his work was to advance "the happiness of mankind," and he continually criticized his predecessors for failing to understand that the real, legitimate, and only goal of the sciences is the "endowment of human life with new inventions and riches." He brought science down from the heavens, including mathematics, which he conceived of as a humble handmaiden to invention. In this utilitarian view of knowledge, Bacon was the chief architect of a new edifice of thought in which resignation was cast out and God assigned to a special room. The name of the building was Progress and Power.

Ironically, Bacon was not himself a scientist, or at least not much of one. He did no pioneering work in any field of research. He did not uncover any new law of nature or generate a single fresh hypothesis. He was not even well informed about the scientific investigations of his own time. And though he prided himself on being the creator of a revolutionary advance in scientific method, posterity has not allowed him this presumption. Indeed, his most famous experiment makes its claim on our attention because Bacon died as a result of it. He and his good friend Dr. Witherborne were taking a coach ride on a wintry day when, seeing snow on the ground, Bacon wondered if flesh might not be preserved in snow, as it is in salt. The two decided to find out at once. They bought a hen, removed its innards, and stuffed the body with snow. Poor Bacon never learned the result of his experiment, because he fell immediately ill from the cold, most probably with bronchitis, and died three days later. For this, he is sometimes regarded as a martyr to experimental science.

But experimental science was not where his greatness lay. Although others of his time were impressed by the effects of practical inventions on the conditions of life, Bacon was the first to think deeply and systematically on the matter. He devoted much of his work to educating men to see the links between invention and progress. In *Novum Organum* he wrote,

> It is well to observe the force and effect and consequences of discoveries. These are to be seen nowhere more conspicuously than in those three which were unknown to the ancients, and of which the origin, though recent, is obscure; namely, printing, gunpowder, and the magnet. For

> these three have changed the whole face and state of things throughout the world; the first in literature, the second in warfare, the third in navigation; whence have followed innumerable changes; insomuch that no empire, no sect, no star seems to have exerted greater power and influence in human affairs than these changes.

In this passage, we can detect some of Bacon's virtues and the source of his great influence. Here is no sleepwalker. He knows full well what technology does to culture and places technological development at the center of his reader's attention. He writes with conviction and verve. He is, after all, among the world's great essayists; Bacon was a master propagandist, who knew well the history of science but saw science not as a record of speculative opinion but as the record of what those opinions had enabled man to do. And he was ceaselessly energetic in trying to convey this idea to his countrymen, if not the world. In the first two books of *Novum Organum,* which consist of 182 aphorisms, Bacon sets out nothing less than a philosophy of science based on the axiom that "the improvement of men's minds and the improvement of his lot are one and the same thing." It is in this work that he denounces the infamous four Idols, which have kept man from gaining power over nature: Idols of the Tribe, which lead us to believe our perceptions are the same as nature's facts; Idols of the Cave, which lead us to mistaken ideas derived from heredity and environment; Idols of the Market-place, which lead us to be deluded by words; and Idols of the Theater, which lead us to the misleading dogmas of the philosophers.

To read Bacon today is to be constantly surprised at his modernity. We are never far from the now familiar notion that science is a source of power and progress. In *The Advancement of Learning,* he even outlines the foundation of a College for Inventors that sounds something like the Massachusetts Institute of Technology. Bacon would have the government provide inventors with allowances for their experiments and for traveling. He would have scholarly journals and international associations. He would encourage full cooperation among scientists, an idea that would have startled Tycho Brahe, Kepler, and Galileo, who used some of their genius to devise ways of concealing their work from one another. Bacon also believed that scientists should be paid well to give public lectures, and that informing the public of the utility of invention was as important as invention itself. In short, he conceived of the scientific enterprise as it is conceived today—organized, financially secure, public, and mankind's best weapon in the struggle to improve his condition and to do so continuously.

As I have said, Bacon is the first man of technocracy, but it was some time before he was joined by the multitude. He died in 1626, and it took another 150 years for European culture to pass to the mentality of the modern world—that is, to technocracy. In doing so, people came to believe that knowledge is power, that humanity is capable of progressing, that poverty is a great evil, and that the life of the average person is as meaningful as any other. It is untrue to say that along the way God died. But any conception of God's design certainly lost much of its power and meaning, and with that loss went the satisfactions of a culture in which moral and intellectual values were integrated. At the same time, we must remember that in the tool-using culture of the older European world, the vast majority of people were peasants, impoverished and powerless. If they believed their afterlife was filled with unending joy, their lives on earth were nonetheless "nasty, brutish and short." As C. P. Snow remarked, the Industrial Revolution of the nineteenth century, which was the fruit of Baconian science, was the only hope for the poor. And if their "true Deity became mechanism," as Thomas Carlyle said, it is probable that by then most people would not have traded their earthly existence for life in a godly, integrated tool-using culture. It didn't matter if they would, since there was little use in lamenting the past. The Western world had become a technocracy from which there could be no turning back. Addressing both those who were exhilarated by technocracy and those who were repulsed by it, Stephen Vincent Benét gave the only advice that made any sense. In *John Brown's Body* he wrote:

> *If you at last must have a word to say,*
> *Say neither, in their way,*
> *"It is a deadly magic and accursed,"*
> *Nor "It is blest," but only "It is here."*

3

From Technocracy to Technopoly

Say only, "It is here." But when did "here" begin? When did Bacon's ideology become a reality? When, to use Siegfried Giedion's phrase, did mechanization take command? To be cautious about it, we might locate the emergence of the first true technocracy in England in the latter half of the eighteenth century—let us say with James Watt's invention of the steam engine in 1765. From that time forward, a decade did not pass without the invention of some significant machinery which, taken together, put an end to medieval "manufacture" (which once meant "to make by hand"). The practical energy and technical skills unleashed at this time changed forever the material and psychic environment of the Western world.

An equally plausible date for the beginnings of technocracy (and, for Americans, easier to remember) is 1776, when Adam Smith's *Wealth of Nations* was published. As Bacon was no scientist, Smith was no inventor. But, like Bacon, he provided a theory that gave conceptual relevance and credibility to the direction in which human enterprise was pointed. Specifically, he justified the transformation from small-scale, personalized, skilled labor to large-scale, impersonal, mechanized production. He not only argued convincingly that money, not land, was the key to wealth, but gave us his famous principle of the self-regulating market. In a technocracy—that is, a society only loosely controlled by social custom and religious tradition and driven by the impulse to invent—an "unseen hand" will eliminate the incompetent and reward those who produce cheaply and well the goods that people want. It was not clear then, and still isn't, whose unseen mind guides the unseen hand, but it is possible (the technocratic industrialists believed) that God could have something to do with it. And if not God, then "human nature," for Adam Smith had named our species "Economic Man," born with an instinct to barter and acquire wealth.

In any case, toward the end of the eighteenth century, technocracy was well underway, especially after Richard Arkwright, a barber by trade, developed the factory system. In his cotton-spinning mills, Arkwright trained workers, mostly children, "to conform to the regular celerity of the machine," and in doing so gave an enormous boost to the growth of modern forms of technocratic capitalism. In 1780, twenty factories were under his control, for which a grateful nation knighted him, and from which an equally grateful son inherited a fortune. Arkwright may fairly be thought of as the first—even archetypal—technocratic capitalist. He exemplified in every particular the type of nineteenth-century entrepreneur to come. As Siegfried Giedion has described him, Arkwright created the first mechanization of production "[in] a hostile environment, without protectors, without government subsidy, but nourished by a relentless utilitarianism that feared no financial risk or danger." By the beginning of the nineteenth century, England was spawning such entrepreneurs in every major city. By 1806, the concept of the power loom, introduced by Edmund Cartwright (a clergyman no less), was revolutionizing the textile industry by eliminating, once and for all, skilled workers, replacing them with workers who merely kept the machines operating.

By 1850, the machine-tool industry was developed—machines to make machines. And beginning in the 1860s, especially in America, a collective fervor for invention took hold of the masses. To quote Giedion again: "Everyone invented, whoever owned an enterprise sought ways and means to make his goods more speedily, more perfectly, and often of improved beauty. Anonymously and inconspicuously the old tools were transformed into modern instruments."[1] Because of their familiarity, it is not necessary to describe in detail all of the inventions of the nineteenth century, including those which gave substance to the phrase "communications revolution": the photograph and telegraph (1830s), rotary-power printing (1840s), the typewriter (1860s), the transatlantic cable (1866), the telephone (1876), motion pictures and wireless telegraphy (1895). Alfred North Whitehead summed it up best when he remarked that the greatest invention of the nineteenth century was the idea of invention itself. We had learned *how* to invent things, and the question of *why* we invent things receded in importance. The idea that if something could be done it should be done was born in the nineteenth century. And along with it, there developed a profound belief in all the principles through which invention succeeds: objectivity, efficiency, expertise, standardization, measurement, and progress. It also came to be believed that the engine of technological progress worked most efficiently when people are conceived of not as children of God or even as citizens but as consumers—that is to say, as markets.

Not everyone agreed, of course, especially with the last notion. In England, William Blake wrote of the "dark Satanic mills" which stripped men of their souls. Matthew Arnold warned that "faith in machinery" was mankind's greatest menace. Carlyle, Ruskin, and William Morris railed against the spiritual degradation brought by industrial progress. In France,

Balzac, Flaubert, and Zola documented in their novels the spiritual emptiness of "Economic man" and the poverty of the acquisitive impulse.

The nineteenth century also saw the emergence of "utopian" communities, of which perhaps the most famous is Robert Owen's experimental community in Scotland called New Lanark. There, he established a model factory community, providing reduced working hours, improved living conditions, and innovative education for the children of workers. In 1824, Owen came to America and founded another utopia at New Harmony, Indiana. Although none of his or other experiments endured, dozens were tried in an effort to reduce the human costs of a technocracy.[2]

We also must not omit mentioning the rise and fall of the much-maligned Luddite Movement. The origin of the term is obscure, some believing that it refers to the actions of a youth named Ludlum who, being told by his father to fix a weaving machine, proceeded instead to destroy it. In any case, between 1811 and 1816, there arose widespread support for workers who bitterly resented the new wage cuts, child labor, and elimination of laws and customs that had once protected skilled workers. Their discontent was expressed through the destruction of machines, mostly in the garment and fabric industry; since then the term "Luddite" has come to mean an almost childish and certainly naïve opposition to technology. But the historical Luddites were neither childish nor naïve. They were people trying desperately to preserve whatever rights, privileges, laws, and customs had given them justice in the older world-view.[3]

They lost. So did all the other nineteenth-century nay-sayers. Copernicus, Kepler, Galileo, and Newton might well have been on their side. Perhaps Bacon as well, for it was not his intention that technology should be a blight or a destroyer. But then, Bacon's greatest deficiency had always been that he was unfamiliar with the legend of Thamus; he understood nothing of the dialectic of technological change, and said little about the negative consequences of technology. Even so, taken as a whole, the rise of technocracy would probably have pleased Bacon, for there can be no disputing that technocracy transformed the face of material civilization, and went far toward relieving what Tocqueville called "the disease of work." And though it is true that technocratic capitalism created slums and alienation, it is also true that such conditions were perceived as an evil that could and should be eradicated; that is to say, technocracies brought into being an increased respect for the average person, whose potential and even convenience became a matter of compelling political interest and urgent social policy. The nineteenth century saw the extension of public education, laid the foundation of the modern labor union, and led to the rapid diffusion of literacy, especially in America, through the development of public libraries and the increased importance of the general-interest magazine. To take only one example of the last point, the list of nineteenth-century contributors to *The Saturday Evening Post,* founded in 1821, included William Cullen Bryant, Harriet Beecher Stowe, James Fenimore Cooper, Ralph Waldo Emerson, Nathaniel Hawthorne, and Edgar Allan Poe—in other words, most of the writers presently included in American Lit. 101. The technocratic culture eroded the line that had made the intellectual interests of educated people inaccessible to the working class, and we may take it as a fact, as George Steiner has remarked, that the period from the French Revolution to World War I marked an oasis of quality in which great literature reached a mass audience.

Something else reached a mass audience as well: political and religious freedom. It would be an inadmissible simplification to claim that the Age of Enlightenment originated solely because of the emerging importance of technology in the eighteenth century, but it is quite clear that the great stress placed on individuality in the economic sphere had an irresistible resonance in the political sphere. In a technocracy, inherited royalty is both irrelevant and absurd. The new royalty was reserved for men like Richard Arkwright, whose origins were low but whose intelligence and daring soared. Those possessed of such gifts could not be denied political power and were prepared to take it if it were not granted. In any case, the revolutionary nature of the new means of production and communication would have naturally generated radical ideas in every realm of human enterprise. Technocracy gave us the idea of progress, and of necessity loosened our bonds with tradition—whether political or spiritual. Technocracy filled the air with the promise of new freedoms and new forms of social organization. Technocracy also speeded up the world. We could get places faster, do things faster, accomplish more in a shorter time. Time, in fact, became an adversary over which technology could triumph. And this meant that there was no time to look back or to contemplate what was being lost. There were empires to build, opportunities to exploit, exciting freedoms to enjoy, especially in America. There, on the wings of technocracy, the United States soared to unprecedented heights as a world power. That Jefferson, Adams, and Madison would have found such a place uncomfortable, perhaps even disagreeable, did not matter. Nor did it matter that there were nineteenth-century American voices—Thoreau, for example—who complained about what was being left behind. The first answer to the complaints was, We leave nothing behind but the chains of a tool-using culture. The second answer was more thoughtful: Technocracy will not overwhelm us. And this was true, to a degree. Technocracy did not entirely destroy the traditions of the social and symbolic worlds. Technocracy subordinated these worlds—yes, even humiliated them—but it did not render them totally ineffectual. In nineteenth-century America, there still existed holy men and the concept of sin. There still existed regional pride, and it was possible to conform to traditional notions of family life. It was possible to respect tradition itself and to find sustenance in ritual and myth. It was possible to believe in social responsibility and the practicality of individual action. It was even possible to believe in common sense and the wisdom of the elderly. It was not easy, but it was possible.

The technocracy that emerged, fully armed, in nineteenth-century America disdained such beliefs, because holy men and sin, grandmothers and families, regional loyalties and two-thousand-year-old traditions, are antagonistic to the techno-cratic way of life. They are a troublesome residue of a tool-using period, a source of criticism of technocracy. They represent a thought-world that stands apart from technocracy and rebukes it—rebukes its language, its impersonality, its fragmentation, its alienation. And so technocracy disdains such a thought-world but, in America, did not and could not destroy it.

We may get a sense of the interplay between technocracy and Old World values in the work of Mark Twain, who was fascinated by the technical accomplishments of the nineteenth century. He said of it that it was "the plainest and sturdiest and infinitely greatest and worthiest of all the centuries the world has seen," and he once congratulated Walt Whitman on having lived in the age that gave the world the beneficial products of coal tar. It is often claimed that he was the first writer regularly to use a typewriter, and he invested (and lost) a good deal of money in new inventions. In his *Life on the Mississippi,* he gives lovingly detailed accounts of industrial development, such as the growth of the cotton mills in Natchez:

> The Rosalie Yarn Mill of Natchez has a capacity of 6000 spindles and 160 looms, and employs 100 hands. The Natchez Cotton Mills Company began operations four years ago in a two-story building of 50 x 190 feet, with 4000 spindles and 128 looms. . . . The mill works 5000

bales of cotton annually and manufactures the best standard quality of brown shirtings and sheetings and drills, turning out 5,000,000 yards of these goods per year.

Twain liked nothing better than to describe the giantism and ingenuity of American industry. But at the same time, the totality of his work is an affirmation of preindustrial values. Personal loyalty, regional tradition, the continuity of family life, the relevance of the tales and wisdom of the elderly are the soul of his work throughout. The story of Huckleberry Finn and Jim making their way to freedom on a raft is nothing less than a celebration of the enduring spirituality of pretechnological man.

If we ask, then, why technocracy did not destroy the world-view of a tool-using culture, we may answer that the fury of industrialism was too new and as yet too limited in scope to alter the needs of inner life or to drive away the language, memories, and social structures of the tool-using past. It was possible to contemplate the wonders of a mechanized cotton mill without believing that tradition was entirely useless. In reviewing nineteenth-century American history, one can hear the groans of religion in crisis, of mythologies under attack, of a politics and education in confusion, but the groans are not yet death-throes. They are the sounds of a culture in pain, and nothing more. The ideas of tool-using cultures were, after all, designed to address questions that still lingered in a technocracy. The citizens of a technocracy knew that science and technology did not provide philosophies by which to live, and they clung to the philosophies of their fathers. They could not convince themselves that religion, as Freud summed it up at the beginning of the twentieth century, is nothing but an obsessional neurosis. Nor could they quite believe, as the new cosmology taught, that the universe is the outcome of accidental collocations of atoms. And they continued to believe, as Mark Twain did, that, for all their dependence on machinery, tools ought still to be their servants, not their masters. They would allow their tools to be presumptuous, aggressive, audacious, impudent servants, but that tools should rise above their servile station was an appalling thought. And though technocracy found no clear place for the human soul, its citizens held to the belief that no increase in material wealth would compensate them for a culture that insulted their self-respect.

And so two opposing world-views—the technological and the traditional—coexisted in uneasy tension. The technological was the stronger, of course, but the traditional was there—still functional, still exerting influence, still too much alive to ignore. This is what we find documented not only in Mark Twain but in the poetry of Walt Whitman, the speeches of Abraham Lincoln, the prose of Thoreau, the philosophy of Emerson, the novels of Hawthorne and Melville, and, most vividly of all, in Alexis de Tocqueville's monumental *Democracy in America.* In a word, two distinct thought-worlds were rubbing against each other in nineteenth-century America.

With the rise of Technopoly, one of those thought-worlds disappears. Technopoly eliminates alternatives to itself in precisely the way Aldous Huxley outlined in *Brave New World.* It does not make them illegal. It does not make them immoral. It does not even make them unpopular. It makes them invisible and therefore irrelevant. And it does so by redefining what we mean by religion, by art, by family, by politics, by history, by truth, by privacy, by intelligence, so that our definitions fit its new requirements. Technopoly, in other words, is totalitarian technocracy.

As I write (in fact, it is the reason why I write), the United States is the only culture to have become a Technopoly. It is a young Technopoly, and we can assume that it wishes not merely to have been the first but to remain the most highly developed. Therefore, it watches with a careful eye Japan and several European nations that are striving to become Technopolies as well.

To give a date to the beginnings of Technopoly in America is an exercise in arbitrariness. It is somewhat like trying to say, precisely, when a coin you have flipped in the air begins its descent. You cannot see the exact moment it stops rising; you know only that it has and is going the other way. Huxley himself identified the emergence of Henry Ford's empire as the decisive moment in the shift from technocracy to Technopoly, which is why in his brave new world time is reckoned as BF (Before Ford) and AF (After Ford).

Because of its drama, I am tempted to cite, as a decisive moment, the famous Scopes "monkey" trial held in Dayton, Tennessee, in the summer of 1925. There, as with Galileo's heresy trial three centuries earlier, two opposing world-views faced each other, toe to toe, in unconcealed conflict. And, as in Galileo's trial, the dispute focused not only on the content of "truth" but also on the appropriate process by which "truth" was to be determined. Scopes' defenders brought forward (or, more accurately, tried to bring forward) all the assumptions and methodological ingenuity of modern science to demonstrate that religious belief can play no role in discovering and understanding the origins of life. William Jennings Bryan and his followers fought passionately to maintain the validity of a belief system that placed the question of origins in the words of their god. In the process, they made themselves appear ridiculous in the eyes of the world. Almost seventy years later, it is not inappropriate to say a word in their behalf: These "fundamentalists" were neither ignorant of nor indifferent to the benefits of science and technology. They had automobiles and electricity and machine-made clothing. They used telegraphy and radio, and among their number were men who could fairly be called reputable scientists. They were eager to share in the largesse of the American technocracy, which is to say they were neither Luddites nor primitives. What wounded them was the assault that science made on the ancient story from which their sense of moral order sprang. They lost, and lost badly. To say, as Bryan did, that he was more interested in the Rock of Ages than the age of rocks was clever and amusing but woefully inadequate. The battle settled the issue, once and for all: in defining truth, the great narrative of inductive science takes precedence over the great narrative of Genesis, and those who do not agree must remain in an intellectual backwater.

Although the Scopes trial has much to recommend it as an expression of the ultimate repudiation of an older world-view, I must let it pass. The trial had more to do with science and faith than technology *as* faith. To find an event that signaled the beginning of a technological theology, we must look to a slightly earlier and less dramatic confrontation. Not unmindful of its value as a pun, I choose what happened in the fall of 1910 as the critical symptom of the onset of Technopoly. From September through November of that year, the Interstate Commerce Commission held hearings on the application of Northeastern railroads for an increase in freight rates to compensate for the higher wages railroad workers had been awarded earlier in the year. The trade association, represented by Louis Brandeis, argued against the application by claiming that the railroads could increase their profits simply by operating more efficiently. To give substance to the argument, Brandeis brought forward witnesses—mostly engineers and industrial managers—who claimed that the railroads could both increase wages and lower their costs by using principles of *scientific management.* Although Frederick W. Taylor was not present at the hearings, his name was frequently invoked as the originator of scientific management, and experts assured the commission that the system developed by Taylor could solve everyone's problem. The commission ultimately ruled against the railroad's application, mostly because it judged that the railroads were making enough money as things were, not because it believed in scientific management. But many people did believe, and the hearings projected Taylor and his system onto the national scene. In the years that followed, attempts were made to apply the principles of the Taylor System in the armed forces, the legal profession, the home, the church, and education. Eventually, Taylor's name and the specifics of his system faded into obscurity, but his ideas about what culture is made of remain the scaffolding of the present-day American Technopoly.

I use this event as a fitting starting point because Taylor's book *The Principles of Scientific Management*, published in 1911, contains the first explicit and formal outline of the assumptions of the thought-world of Technopoly. These include the beliefs that the primary, if not the only, goal of human labor and thought is efficiency; that technical calculation is in all respects superior to human judgment; that in fact human judgment cannot be trusted, because it is plagued by laxity, ambiguity, and unnecessary complexity; that subjectivity is an obstacle to clear thinking; that what cannot be measured either does not exist or is of no value; and that the affairs of citizens are best guided and conducted by experts. In fairness to Taylor (who did not invent the term "scientific management" and who used it reluctantly), it should be noted that his system was originally devised to apply only to industrial production. His intention was to make a science of the industrial workplace, which would not only increase profits but also result in higher wages, shorter hours, and better working conditions for laborers. In his system, which included "time and motion studies," the judgment of individual workers was replaced by laws, rules, and principles of the "science" of their job. This did mean, of course, that workers would have to abandon any traditional rules of thumb they were accustomed to using; in fact, workers were relieved of any responsibility to think at all. The system would do their thinking for them. That is crucial, because it led to the idea that technique of any kind can do our thinking for us, which is among the basic principles of Technopoly.

The assumptions that underlay the principles of scientific management did not spring, all at once, from the originality of Taylor's mind. They were incubated and nurtured in the technocracies of the eighteenth and nineteenth centuries. And a fair argument can be made that the origins of Technopoly are to be found in the thought of the famous nineteenth-century French philosopher Auguste Comte, who founded both positivism and sociology in an effort to construct a science of society. Comte's arguments for the unreality of anything that could not be seen and measured certainly laid the foundation for the future conception of human beings as objects. But in a technocracy, such ideas exist only as by-products of the increased role of technology. Technocracies are concerned to invent machinery. That people's lives are changed by machinery is taken as a matter of course, and that people must sometimes be treated as if they were machinery is considered a necessary and unfortunate condition of technological development. But in technocracies, such a condition is not held to be a philosophy of culture. Technocracy does not have as its aim a grand reductionism in which human life must find its meaning in machinery and technique. Technopoly does. In the work of Frederick Taylor we have, I believe, the first clear statement of the idea that society is best served when human beings are placed at the disposal of their techniques and technology, that human beings are, in a sense, worth less than their machinery. He and his followers described exactly what this means, and hailed their discovery as the beginnings of a brave new world.

Why did Technopoly—the submission of all forms of cultural life to the sovereignty of technique and technology—find fertile ground on American soil? There are four interrelated reasons for the rise of Technopoly in America, why it emerged in America first, and why it has been permitted to flourish. As it happens, all of these have been written about extensively in many contexts and are well known. The first concerns what is usually called the American character, the relevant aspeet of which Tocqueville described in the early nineteenth century. "The American lives in a land of wonders," he wrote; "everything around him is in constant movement, and every movement seems an advance. Consequently, in his mind the idea of newness is closely linked with that of improvement. Nowhere does he see any limit placed by nature to human endeavor; in his eyes something that does not exist is just something that has not been tried."[4]

This feature of the American ethos is plain to everyone who has studied American culture, although there are wide variations in the explanation of it. Some attribute it to the immigrant nature of the population; some to the frontier mentality; some to the abundant natural resources of a singularly blessed land and the unlimited opportunities of a new continent; some to the unprecedented political and religious freedom afforded the average person; some to all of these factors and more. It is enough to say here that the American distrust of constraints—one might even say the American skepticism toward culture itself—offered encouragement to radical and thoughtless technological intrusions.

Second, and inextricably related to the first, is the genius and audacity of American capitalists of the late nineteenth and early twentieth centuries, men who were quicker and more focused than those of other nations in exploiting the economic possibilities of new technologies. Among them are Samuel Morse, Alexander Graham Bell, Thomas Edison, John D. Rockefeller, John Jacob Astor, Henry Ford, Andrew Carnegie, and many others, some of whom were known as Robber Barons. What they were robbing—it is clearer now than it was then—was America's past, for their essential idea was that nothing is so much worth preserving that it should stand in the way of technological innovation. These were the men who created the twentieth century, and they achieved wealth, prestige, and power that would have amazed even Richard Arkwright. Their greatest achievement was in convincing their countrymen that the future need have no connection to the past.

Third, the success of twentieth-century technology in providing Americans with convenience, comfort, speed, hygiene, and abundance was so obvious and promising that there seemed no reason to look for any other sources of fulfillment or creativity or purpose. To every Old World belief, habit, or tradition, there was and still is a technological alternative. To prayer, the alternative is penicillin; to family roots, the alternative is mobility; to reading, the alternative is television; to restraint, the alternative is immediate gratification; to sin, the alternative is psychotherapy; to political ideology, the alternative is popular appeal established through scientific polling. There is even an alternative to the painful riddle of death, as Freud called it. The riddle may be postponed through longer life, and then perhaps solved altogether by cryogenics. At least, no one can easily think of a reason why not.

As the spectacular triumphs of technology mounted, something else was happening: old sources of belief came under siege. Nietzsche announced that God was dead. Darwin didn't go as far but did make it clear that, if we were children of God, we had come to be so through a much longer and less dignified route than we had imagined, and that in the process we had picked up some strange and unseemly relatives. Marx argued that history had its own agenda and was taking us where it must, irrespective of our wishes. Freud taught that we had no understanding of our deepest needs and could not trust our traditional ways of reasoning to uncover them. John Watson, the founder of behaviorism, showed that free will was an illusion and that our behavior, in the end, was not unlike that of pigeons. And Einstein and his colleagues told us that there were no absolute means of judging anything in any case, that everything was relative. The thrust of a century of scholarship had the effect of making us lose confidence in our belief systems and therefore in ourselves. Amid the conceptual debris, there remained one sure thing to believe in—technology. Whatever else may be denied or compromised, it is clear that airplanes do fly, antibiotics do cure, radios do speak, and, as we know now, computers do calculate and never make mistakes—only faulty humans do (which is what Frederick Taylor was trying to tell us all along).

For these well-known reasons, Americans were better prepared to undertake the creation of a Technopoly than anyone else. But its full flowering depended on still another set of conditions, less visible and therefore less well known. These conditions provided the background, the context in which the American distrust of constraints, the exploitative genius of its captains of industry, the successes of technology, and the devaluation of traditional beliefs took on the exaggerated significance that pushed technocracy in America over into Technopoly. That context is explored in the following chapter, which I call "The Improbable World."

4

The Improbable World

Although it is clear that "social science" is a vigorous ally of Technopoly and must therefore be regarded with a hostile eye, I occasionally pay my respects to its bloated eminence by inflicting a small experiment on some of my colleagues. Like many other social-science experiments, this one is based on deceit and exploitation, and I must rely on the reader's sense of whimsy to allow its point to come through.

The experiment is best conducted in the morning when I see a colleague who appears not to be in possession of a copy of *The New York Times.* "Did you read the *Times* this morning?" I ask. If my colleague says, "Yes," there is no experiment that day. But if the answer is "No," the experiment can proceed. "You ought to check out Section C today," I say. "There's a fascinating article about a study done at the University of Minnesota." "Really? What's it about?" is the usual reply. The choices at this point are almost endless, but there are two that produce rich results. The first: "Well, they did this study to find out what foods are best to eat for losing weight, and it turns out that a normal diet supplemented by chocolate eclairs eaten three times a day is the best approach. It seems that there's some special nutrient in the eclairs—encomial dyoxin—that actually uses up calories at an incredible rate."

The second changes the theme and, from the start, the university: "The neurophysiologists at Johns Hopkins have uncovered a connection between jogging and reduced intelligence. They tested more than twelve hundred people over a period of five years, and found that as the number of hours people jogged increased there was a statistically significant decrease in their intelligence. They don't know exactly why, but there it is."

My role in the experiment, of course, is to report something quite ridiculous—one might say, beyond belief. If I play my role with a sense of decorum and collegial intimacy, I can achieve results worth reporting: about two-thirds of the victims will believe or at least not wholly *disbelieve* what I have told them. Sometimes they say, "Really? Is that possible?" Sometimes they do a double-take and reply, "*Where'd* you say that study was done?" And sometimes they say, "You know, I've *heard* something like that." I should add that for reasons that are probably worth exploring I get the clearest cases of credulity when I use the University of Minnesota and Johns Hopkins as my sources of authority; Stanford and MIT give only fair results.

There are several conclusions that might be drawn from these results, one of which was expressed by H. L. Mencken fifty years ago, when he said that there is no idea so stupid that you can't find a professor who will believe it. This is more an accusation than an explanation, although there is probably something to it. (I have, however, tried this experiment on nonprofessors as well, and get roughly the same results.) Another possible conclusion was expressed by George Bernard Shaw, also about fifty years ago, when he wrote that the average person today is about as credulous as was the average person in the Middle Ages. In the Middle Ages, people believed in the authority of their religion, no matter what. Today, we believe in the authority of our science, no matter what.

However, there is still another possibility, related to Shaw's point but off at a right angle to it. It is, in any case, more relevant to understanding the sustaining power of Technopoly. I mean that the world we live in is very nearly incomprehensible to most of us. There is almost no fact, whether actual or imagined, that will surprise us for very long, since we have no comprehensive and consistent picture of the world that would make the fact appear as an unacceptable contradiction. We believe because there is no reason not to believe. And I assume that the reader does not need the evidence of my comic excursion into the suburbs of social science to recognize this. Abetted by a form of education that in itself has been emptied of any coherent world-view, Technopoly deprives us of the social, political, historical, metaphysical, logical, or spiritual bases for knowing what is beyond belief.

That is especially the case with technical facts. Since this book is filled with a variety of facts, I would hardly wish to shake confidence in them by trying my experiment on the reader. But if I informed you that the paper on which this book is printed was made by a special process which uses the skin of a pickled herring, on what grounds would you dispute me? For all you know—indeed, for all *I* know—the skin of a pickled herring could have made this paper. And if the facts were confirmed by an industrial chemist who described to us some incomprehensible process by which it was done (employing, of course, encomial dyoxin), we might both believe it. Or not wholly disbelieve it, since the ways of technology, like the ways of God, are awesome and mysterious.

Perhaps I can get a bit closer to the point with an analogy. If you open a brand-new deck of cards and start turning the cards over, one by one, you can get a pretty firm idea of what their order is. After you have gone from the ace of spades through to the nine of spades, you expect a ten of spades to come up next. And if the three of diamonds appears, you are surprised and wonder what kind of deck of cards this is. But if I give you a deck that had been shuffled twenty times and then ask you to turn the cards over, you do not expect any card in particular—a three of diamonds would be just as likely as a ten of spades. Having no expectation of a pattern, no basis for assuming a given order, you have no reason to react with incredulity or even surprise to whatever card turns up.

The belief system of a tool-using culture is rather like a brand-new deck of cards. Whether it is a culture of technological simplicity or sophistication, there always exists a more or less comprehensive, ordered world-view, resting on a set of metaphysical or theological assumptions. Ordinary men and women might not clearly grasp how the harsh realities of their lives fit into the grand and benevolent design of the universe, but they have no doubt that there *is* such a design, and their priests and shamans are well able, by deduction from a handful of principles, to make it, if not wholly rational, at least coherent. The medieval period was a particularly clear example of this point. How comforting it must have been to have a priest explain the meaning of the death of a loved one, of an accident, or of a piece of good fortune. To live in a world in which there were no random events—in which everything was, in theory, comprehensible; in which every act of nature was infused with meaning—is an irreplaceable gift of theology. The role of the church in premodern Europe was to keep the deck of cards in reasonable order, which is why Cardinal Bellarmine and other prelates tried to prevent Galileo from shuffling the deck. As we know, they could not, and with the emergence of technocracies moral and intellectual coherence began to unravel.

What was being lost was not immediately apparent. The decline of the great narrative of the Bible, which had provided answers to both fundamental and practical questions, was accompanied by the rise of the great narrative of Progress. The faith of those who believed in Progress was based on the assumption that one could discern a purpose to the human enterprise, even without the theological scaffolding that supported the Christian edifice of belief. Science and technology were the chief instruments of Progress, and in their accumulation of reliable information about nature they would bring ignorance, superstition, and suffering to an end. As it turned out, technocracies did not disappoint Progress. In sanitation, pharmacology, transportation, production, and communication, spectacular improvements were made possible by a Niagara of information generated by just such institutions as Francis Bacon had imagined. Technocracy was fueled by information—about the structure of nature as well as the structure of the human soul.

But the genie that came out of the bottle proclaiming that information was the new god of culture was a deceiver. It solved the problem of information scarcity, the disadvantages of which were obvious. But it gave no warning about the dangers of information glut, the disadvantages of which were not seen so clearly. The long-range result—information chaos—has produced a culture somewhat like the shuffled deck of cards I referred to. And what is strange is that so few have noticed, or if they have noticed fail to recognize the source of their distress. You need only ask yourself, What is the problem in the Middle East, or South Africa, or Northern Ireland? Is it lack of information that keeps these conflicts at fever pitch? Is it lack of information about how to grow food that keeps millions at starvation levels? Is it lack of information that brings soaring crime rates and physical decay to our cities? Is it lack of information that leads to high divorce rates and keeps the beds of mental institutions filled to overflowing?

The fact is, there are very few political, social, and especially personal problems that arise because of insufficient information.

Nonetheless, as incomprehensible problems mount, as the concept of progress fades, as meaning itself becomes suspect, the Technopolist stands firm in believing that what the world needs is yet more information. It is like the joke about the man who complains that the food he is being served in a restaurant is inedible and also that the portions are too small. But, of course, what we are dealing with here is no joke. Attend any conference on telecommunications or computer technology, and you will be attending a celebration of innovative machinery that generates, stores, and distributes more information, more conveniently, at greater speeds than ever before. To the question "What problem does the information solve?" the answer is usually "How to generate, store, and distribute more information, more conveniently, at greater speeds than ever before." This is the elevation of information to a metaphysical status: information as both the means and end of human creativity. In Technopoly, we are driven to fill our lives with the quest to "access" information. For what purpose or with what limitations, it is not for us to ask; and we are not accustomed to asking, since the problem is unprecedented. The world has never before been confronted with information glut and has hardly had time to reflect on its consequences.

As with so many of the features of all that is modern, the origins of information glut can be traced many centuries back. Nothing could be more misleading than the claim that computer technology introduced the age of information. The printing press began that age in the early sixteenth century.[1] Forty years after Gutenberg converted an old wine press into a printing machine with movable type, there were presses in 110 cities in six different countries. Fifty years after the press was invented, more than eight million books had been printed, almost all of them filled with information that had previously been unavailable to the average person. There were books on law, agriculture, politics, exploration, metallurgy, botany, linguistics, pediatrics, and even good manners. There were also assorted guides and manuals; the world of commerce rapidly became a world of printed paper through the widespread use of contracts, deeds, promissory notes, and maps. (Not surprisingly, in a culture in which information was becoming standardized and repeatable, mapmakers began to exclude "paradise" from their charts on the grounds that its location was too uncertain.)

So much new information, of so many diverse types, was generated that printers could no longer use the scribal manuscript as their model of a book. By the mid-sixteenth century, printers began to experiment with new formats, among the most important innovations being the use of Arabic numerals to number pages. (The first known example of such pagination is Johann Froben's first edition of Erasmus' New Testament, printed in 1516.) Pagination led inevitably to more accurate indexing, annotation, and cross-referencing, which in turn was accompanied by innovations in punctuation marks, section heads, paragraphing, title-paging, and running heads. By the end of the sixteenth century, the machine-made book had a typographic form and a look comparable to books of today.

All of this is worth mentioning because innovations in the format of the machine-made book were an attempt to control the flow of information, to organize it by establishing priorities and by giving it sequence. Very early on, it was understood that the printed book had created an information crisis and that something needed to be done to maintain a measure of control. The altered form of the book was one means. Another was the modern school, which took shape in the seventeenth century. In 1480, before the information explosion, there were thirty-four schools in all of England. By 1660, there were 444, one school for every twelve square miles. There were several reasons for the rapid growth of the common school, but none was more obvious than that it was a necessary response to the anxieties and confusion aroused by information on the loose. The invention of what is called a curriculum was a logical step toward organizing, limiting, and discriminating among available sources of information. Schools became technocracy's first secular bureaucracies, structures for legitimizing some parts of the flow of information and discrediting other parts. Schools were, in short, a means of governing the ecology of information.

With the rise of technocracies, information became a more serious problem than ever, and several methods of controlling information had to be invented. For a richly detailed account of what those methods were, I refer the reader to James Beniger's *The Control Revolution,* which is among the three or four most important books we have on the subject of the relation of information to culture. In the next chapter, I have relied to a considerable degree on *The Control Revolution* in my discussion of the breakdown of the control mechanisms, but here I must note that most of the methods by which technocracies have hoped to keep information from running amok are now dysfunctional.

Indeed, one way of defining a Technopoly is to say that its information immune system is inoperable. Technopoly is a form of cultural AIDS, which I here use as an acronym for Anti-Information Deficiency Syndrome. This is why it is possible to say almost anything without contradiction provided you begin your utterance with the words "A study has shown . . ." or "Scientists now tell us that . . ." More important, it is why in a Technopoly there can be no transcendent sense of purpose or meaning, no cultural coherence. Information is dangerous when it has no place to go, when there is no theory to which it applies, no pattern in which it fits, when there is no higher purpose that it serves. Alfred North Whitehead called such information "inert," but that metaphor is too passive. Information without regulation can be lethal. It is necessary, then, to describe briefly the technological conditions that led to such a grim state of affairs.

If the telescope was the eye that gave access to a world of new facts and new methods of obtaining them, then the printing press was the larynx. The press not only created new sources of data collection but vastly increased communication among scientists on a continent-wide basis. The movement toward standardization of scientific discourse resulted, for example, in uniform mathematical symbols, including the replacement of Roman with Arabic numerals. Galileo's and Kepler's reference to mathematics as the language or alphabet of nature could be made with assurance that other scientists could speak and understand that language. Standardization largely eliminated ambiguity in texts and reduced error in diagrams, charts, and visual aids. Printing brought an end to the alchemists' secrets by making science into a public enterprise. And not only for scientists: printing led to the popularization of scientific ideas through the use of vernaculars. Although some scientists—Harvey, for example—insisted on writing in Latin, many others (Bacon, of course) eagerly employed the vernacular in an effort to convey the new spirit and methods of scientific philosophy. When we consider that Vesalius, Brahe, Bacon, Galileo, Kepler, Harvey, and Descartes were all born in the sixteenth century, we can begin to grasp the relationship between the growth of science and the printing press, which is to say, the press announced the advent of science, publicized it, encouraged it, and codified it.

As is known, the press did the same for what is now called Protestantism. Martin Luther's reliance on printed pamphlets and books as a means of religious propaganda is well documented, as is his own acknowledgment of the importance of print to his mission. And yet, for all of Luther's astuteness about printing, even he was surprised on occasion by the unsuspected powers of the press. "It is a mystery to me," he wrote in a letter to the Pope, "how my theses . . . were spread to so many places. They were meant exclusively for our academic circle here.

. . . They were written in such a language that the common people could hardly understand them." What Luther overlooked was the sheer *portability* of printed books. Although his theses were written in academic Latin, they were easily transported throughout Germany and other countries by printers who just as easily had them translated into vernaculars.

Without going any further into the details of the impact of print on medieval thought, all of which are lucidly presented in Elizabeth Eisenstein's *The Printing Press as an Agent of Change*, I will instead merely assert the obvious point: By the beginning of the seventeenth century, an entirely new information environment had been created by print. Astronomy, anatomy, and physics were accessible to anyone who could read. New forms of literature, such as the novel and personal essays, were available. Vernacular Bibles turned the Word of God into the words of God, since God became an Englishman or a German or a Frenchman, depending on the language in which His words were revealed. Practical knowledge about machines, agriculture, and medicine was widely dispersed. Commercial documents gave new form and vigorous impetus to entrepreneurial adventures. And, of course, printing vastly enhanced the importance of individuality.

Vitalized by such an information explosion, Western culture set itself upon a course which made technocracies possible. And then something quite unexpected happened; in a word, nothing. From the early seventeenth century, when Western culture undertook to reorganize itself to accommodate the printing press, until the mid-nineteenth century, no significant technologies were introduced that altered the *form, volume,* or *speed* of information. As a consequence, Western culture had more than two hundred years to accustom itself to the new information conditions created by the press. It developed new institutions, such as the school and representative government. It developed new conceptions of knowledge and intelligence, and a heightened respect for reason and privacy. It developed new forms of economic activity, such as mechanized production and corporate capitalism, and even gave articulate expression to the possibilities of a humane socialism. New forms of public discourse came into being through newspapers, pamphlets, broadsides, and books. It is no wonder that the eighteenth century gave us our standard of excellence in the use of reason, as exemplified in the work of Goethe, Voltaire, Diderot, Kant, Hume, Adam Smith, Edmund Burke, Vico, Edward Gibbon, and, of course, Jefferson, Madison, Franklin, Adams, Hamilton, and Thomas Paine. I weight the list with America's "Founding Fathers" because technocratic-typographic America was the first nation ever to be *argued* into existence *in print.* Paine's *Common Sense* and *The Rights of Man,* Jefferson's Declaration of Independence, and the *Federalist Papers* were written and printed efforts to make the American experiment appear reasonable to the people, which to the eighteenth-century mind was both necessary and sufficient. To any people whose politics were the politics of the printed page, as Tocqueville said of America, reason and printing were inseparable. We need not hesitate to claim that the First Amendment to the United States Constitution stands as a monument to the ideological biases of print. It says: "Congress shall make no law respecting the establishment of religion, or prohibiting the free exercise thereof; or abridging freedom of speech or of the press; or of the right of the people peaceably to assemble, and to petition the government for a redress of grievances." In these forty-five words we may find the fundamental values of the literate, reasoning mind as fostered by the print revolution: a belief in privacy, individuality, intellectual freedom, open criticism, and community action.

Equally important is that the words of that amendment presume and insist on a public that not only has access to information but has control over it, a people who know how to use information in their own interests. There is not a single line written by Jefferson, Adams, Paine, Hamilton, or Franklin that does not take for granted that when information is made available to citizens they are capable of managing it. This is not to say that the Founding Fathers believed information could not be false, misleading, or irrelevant. But they believed that the marketplace of information and ideas was sufficiently ordered so that citizens could make sense of what they read and heard and, through reason, judge its usefulness to their lives. Jefferson's proposals for education, Paine's arguments for self-governance, Franklin's arrangements for community affairs assume coherent, commonly shared principles that allow us to debate such questions as: What are the responsibilities of citizens? What is the nature of education? What constitutes human progress? What are the limitations of social structures?

The presumed close connection among information, reason, and usefulness began to lose its legitimacy toward the mid-nineteenth century with the invention of the telegraph. Prior to the telegraph, information could be moved only as fast as a train could travel: about thirty-five miles per hour. Prior to the telegraph, information was sought as part of the process of understanding and solving particular problems. Prior to the telegraph, information tended to be of local interest. Telegraphy changed all of this, and instigated the second stage of the information revolution. The telegraph removed space as an inevitable constraint on the movement of information, and, for the first time, transportation and communication were disengaged from each other. In the United States, the telegraph erased state lines, collapsed regions, and, by wrapping the continent in an information grid, created the possibility of a unified nation-state. But more than this, telegraphy created the idea of context-free information—that is, the idea that the value of information need not be tied to any function it might serve in social and political decision-making and action. The telegraph made information into a commodity, a "thing" that could be bought and sold irrespective of its uses or meaning.[2]

But it did not do so alone. The potential of the telegraph to transform information into a commodity might never have been realized except for its partnership with the penny press, which was the first institution to grasp the significance of the annihilation of space and the saleability of irrelevant information. In fact, the first known use of the telegraph by a newspaper occurred *one day* after Samuel Morse gave his historic demonstration of the telegraph's workability. Using the same Washington-to-Baltimore line Morse had constructed, the Baltimore *Patriot* gave its readers information about action taken by the House of Representatives on the Oregon issue. The paper concluded its report by noting, ". . . we are thus enabled to give our readers information from Washington up to two o'clock. This is indeed the annihilation of space." Within two years of this announcement, the fortunes of newspapers came to depend not on the quality or utility of the news they provided but on how much, from what distances, and at what speed.

And, one must add, with how many photographs. For, as it happened, photography was invented at approximately the same time as telegraphy, and initiated the third stage of the information revolution. Daniel Boorstin has called it "the graphic revolution," because the photograph and other iconographs brought on a massive intrusion of images into the symbolic environment: photographs, prints, posters, drawings, advertisements. The new imagery, with photography at its forefront, did not merely function as a supplement to language but tended to replace it as our dominant means for construing, understanding, and testing reality. By the end of the nineteenth century, advertisers and newspapermen had discovered that a picture was worth not only a thousand words but, in terms of sales, many thousands of dollars.

As the twentieth century began, the amount of information available through words and pictures grew exponentially. With telegraphy and photography leading the way, a new definition of information came into being. Here was information that rejected the necessity of interconnectedness, proceeded without context, argued for instancy against historical continuity, and offered fascination in place of complexity and coherence. And then, with Western culture gasping for breath, the fourth stage of the information revolution occurred, broadcasting. And then the fifth, computer technology. Each of these brought with it new forms of information, unprecedented amounts of it, and increased speeds (if virtual instancy can be increased).

What is our situation today? In the United States, we have 260,000 billboards; 11,520 newspapers; 11,556 periodicals; 27,000 video outlets for renting video tapes; more than 500 million radios; and more than 100 million computers. Ninety-eight percent of American homes have a television set; more than half our homes have more than one. There are 40,000 new book titles published every year (300,000 worldwide), and every day in America 41 million photographs are taken. And if this is not enough, more than 60 billion pieces of junk mail (thanks to computer technology) find their way into our mailboxes every year.

From millions of sources all over the globe, through every possible channel and medium—light waves, airwaves, ticker tapes, computer banks, telephone wires, television cables, satellites, printing presses—information pours in. Behind it, in every imaginable form of storage—on paper, on video and audio tape, on discs, film, and silicon chips—is an ever greater volume of information waiting to be retrieved. Like the Sorcerer's Apprentice, we are awash in information. And all the sorcerer has left us is a broom. Information has become a form of garbage, not only incapable of answering the most fundamental human questions but barely useful in providing coherent direction to the solution of even mundane problems. To say it still another way: The milieu in which Technopoly flourishes is one in which the tie between information and human purpose has been severed, i.e., information appears indiscriminately, directed at no one in particular, in enormous volume and at high speeds, and disconnected from theory, meaning, or purpose.

All of this has called into being a new world. I have referred to it elsewhere as a peek-a-boo world, where now this event, now that, pops into view for a moment, then vanishes again. It is an improbable world. It is a world in which the idea of human progress, as Bacon expressed it, has been replaced by the idea of technological progress. The aim is not to reduce ignorance, superstition, and suffering but to accommodate ourselves to the requirements of new technologies. We tell ourselves, of course, that such accommodations will lead to a better life, but that is only the rhetorical residue of a vanishing technocracy. We are a culture consuming itself with information, and many of us do not even wonder how to control the process. We proceed under the assumption that information is our friend, believing that cultures may suffer grievously from a lack of information, which, of course, they do. It is only now beginning to be understood that cultures may also suffer grievously from information glut, information without meaning, information without control mechanisms.

5

The Broken Defenses

Technopoly is a state of culture. It is also a state of mind. It consists in the deification of technology, which means that the culture seeks its authorization in technology, finds its satisfactions in technology, and takes its orders from technology. This requires the development of a new kind of social order, and of necessity leads to the rapid dissolution of much that is associated with traditional beliefs. Those who feel most comfortable in Technopoly are those who are convinced that technical progress is humanity's supreme achievement and the instrument by which our most profound dilemmas may be solved. They also believe that information is an unmixed blessing, which through its continued and uncontrolled production and dissemination offers increased freedom, creativity, and peace of mind. The fact that information does none of these things—but quite the opposite—seems to change few opinions, for such unwavering beliefs are an inevitable product of the structure of Technopoly. In particular, Technopoly flourishes when the defenses against information break down.

The relationship between information and the mechanisms for its control is fairly simple to describe: Technology increases the available supply of information. As the supply is increased, control mechanisms are strained. Additional control mechanisms are needed to cope with new information. When additional control mechanisms are themselves technical, they in turn further increase the supply of information. When the supply of information is no longer controllable, a general breakdown in psychic tranquillity and social purpose occurs. Without defenses, people have no way of finding meaning in their experiences, lose their capacity to remember, and have difficulty imagining reasonable futures.

One way of defining Technopoly, then, is to say it is what happens to society when the defenses against information glut have broken down. It is what happens when institutional life becomes inadequate to cope with too much information. It is what happens when a culture, overcome by information generated by technology, tries to employ technology itself as a means of providing clear direction and humane purpose. The effort is mostly doomed to failure. Though it is sometimes possible to use a disease as a cure for itself, this occurs only when we are fully aware of the processes by which disease is normally held in check. My purpose here is to describe the defenses that in principle are available and to suggest how they have become dysfunctional.

The dangers of information on the loose may be understood by the analogy I suggested earlier with an individual's biological immune system, which serves as a defense against the uncontrolled growth of cells. Cellular growth is, of course, a normal process without which organic life cannot survive. But without a well-functioning immune system, an organism cannot manage cellular growth. It becomes disordered and destroys the delicate interconnectedness of essential organs. An immune system, in short, destroys unwanted cells. All societies have institutions and techniques that function as does a biological immune system. Their purpose is to maintain a balance between the old and the new, between novelty and tradition, between meaning and conceptual disorder, and they do so by "destroying" unwanted information.

I must emphasize that social institutions of all kinds function as control mechanisms. This is important to say, because most writers on the subject of social institutions (especially sociologists) do not grasp the idea that any decline in the force of institutions makes people vulnerable to information chaos.[1] To say that life is destabilized by weakened institutions is merely to say that information loses its use and therefore becomes a source of confusion rather than coherence.

Social institutions sometimes do their work simply by denying people access to information, but principally by directing how much weight and, therefore, value one must give to information. Social institutions are concerned with the *meaning* of information and can be quite rigorous in enforcing standards of admission. Take as a simple example a court of law. Almost all rules for the presentation of evidence and for the conduct of those who participate in a trial are designed to limit the amount of information that is allowed entry into the system. In our system, a judge disallows "hearsay" or personal opinion as evidence except under strictly controlled circumstances, spectators are forbidden to express their feelings, a defendant's previous convictions may not be mentioned, juries are not allowed to hear arguments over the admissibility of evidence—these are instances of information control. The rules on which such control is based derive from a theory of justice that defines what information may be considered relevant and, especially, what information must be considered irrelevant. The theory may be deemed flawed in some respects—lawyers, for example, may disagree over the rules governing the flow of information—but no one disputes that information must be regulated in some manner. In even the simplest law case, thousands of events may have had a bearing on the dispute, and it is well understood that, if they were all permitted entry, there could be no theory of due process, trials would have no end, law itself would be reduced to meaninglessness. In short, the rule of law is concerned with the "destruction" of information.

It is worth mentioning here that, although legal theory has been taxed to the limit by new information from diverse sources—biology, psychology, and sociology, among them—the rules governing relevance have remained fairly stable. This may account for Americans' overuse of the courts as a means of finding coherence and stability. As other institutions become unusable as mechanisms for the control of wanton information, the courts stand as a final arbiter of truth. For how long, no one knows.

I have previously referred to the school as a mechanism for information control. What its standards are can usually be found in a curriculum or, with even more clarity, in a course catalogue. A college catalogue lists courses, subjects, and fields of study that, taken together, amount to a certified statement of what a serious student ought to think about. More to the point, in what is omitted from a catalogue, we may learn what a serious student ought *not* to think about. A college catalogue, in other words, is a formal description of an information management program; it defines and categorizes knowledge, and in so doing systematically excludes, demeans, labels as trivial—in a word, disregards certain kinds of information. That is why it "makes sense" (or, more accurately, used to make sense). By what it includes/excludes it reflects a theory of the purpose and meaning of education. In the university where I teach, you will not find courses in astrology or dianetics or creationism. There is, of course, much available information about these subjects, but the theory of education that sustains the university does not allow such information entry into the formal structure of its courses.

Professors and students are denied the opportunity to focus their attention on it, and are encouraged to proceed as if it did not exist. In this way, the university gives expression to its idea of what constitutes legitimate knowledge. At the present time, some accept this idea and some do not, and the resulting controversy weakens the university's function as an information control center.

The clearest symptom of the breakdown of the curriculum is found in the concept of "cultural literacy," which has been put forward as an organizing principle and has attracted the serious attention of many educators.[2] If one is culturally literate, the idea goes, one should master a certain list of thousands of names, places, dates, and aphorisms; these are supposed to make up the content of the literate American's mind. But, as I will seek to demonstrate in the final chapter, cultural literacy is not an organizing principle at all; it represents, in fact, a case of calling the disease the cure. The point to be stressed here is that any educational institution, if it is to function well in the management of information, must have a theory about its purpose and meaning, must have the means to give clear expression to its theory, and must do so, to a large extent, by excluding information.

As another example, consider the family. As it developed in Europe in the late eighteenth century, its theory included the premise that individuals need emotional protection from a cold and competitive society. The family became, as Christopher Lasch calls it, a haven in a heartless world.[3] Its program included (I quote Lasch here) preserving "separatist religious traditions, alien languages and dialects, local lore and other traditions." To do this, the family was required to take charge of the socialization of children; the family became a structure, albeit an informal one, for the management of information. It controlled what "secrets" of adult life would be allowed entry and what "secrets" would not. There may be readers who can remember when in the presence of children adults avoided using certain words and did not discuss certain topics whose details and ramifications were considered unsuitable for children to know. A family that does not or cannot control the information environment of its children is barely a family at all, and may lay claim to the name only by virtue of the fact that its members share biological information through DNA. In fact, in many societies a family was just that—a group connected by genetic information, itself controlled through the careful planning of marriages. In the West, the family as an institution for the management of nonbiological information began with the ascendance of print. As books on every conceivable subject become available, parents were forced into the roles of guardians, protectors, nurturers, and arbiters of taste and rectitude. Their function was to define what it means to be a child by excluding from the family's domain information that would undermine its purpose. That the family can no longer do this is, I believe, obvious to everyone.

Courts of law, the school, and the family are only three of several control institutions that serve as part of a culture's information immune system. The political party is another. As a young man growing up in a Democratic household, I was provided with clear instructions on what value to assign to political events and commentary. The instructions did not require explicit statement. They followed logically from theory, which was, as I remember it, as follows: Because people need protection, they must align themselves with a political organization. The Democratic Party was entitled to our loyalty because it represented the social and economic interests of the working class, of which our family, relatives, and neighbors were members (except for one uncle who, though a truck driver, consistently voted Republican and was therefore thought to be either stupid or crazy). The Republican Party represented the interests of the rich, who, by definition, had no concern for us.

The theory gave clarity to our perceptions and a standard by which to judge the significance of information. The general principle was that information provided by Democrats was always to be taken seriously and, in all probability, was both true and useful (except if it came from Southern Democrats, who were helpful in electing presidents but were otherwise never to be taken seriously because of their special theory of race). Information provided by Republicans was rubbish and was useful only to the extent that it confirmed how self-serving Republicans were.

I am not prepared to argue here that the theory was correct, but to the accusation that it was an oversimplification I would reply that all theories are oversimplifications, or at least lead to oversimplification. The rule of law is an oversimplification. A curriculum is an oversimplification. So is a family's conception of a child. That is the function of theories—to oversimplify, and thus to assist believers in organizing, weighting, and excluding information. Therein lies the power of theories. Their weakness is that precisely because they oversimplify, they are vulnerable to attack by new information. When there is too much information to sustain *any* theory, information becomes essentially meaningless.

The most imposing institutions for the control of information are religion and the state. They do their work in a somewhat more abstract way than do courts, schools, families, or political parties. They manage information through the creation of myths and stories that express theories about fundamental questions: why are we here, where have we come from, and where are we headed? I have already alluded to the comprehensive theological narrative of the medieval European world and how its great explanatory power contributed to a sense of well-being and coherence. Perhaps I have not stressed enough the extent to which the Bible also served as an information control mechanism, especially in the moral domain. The Bible gives manifold instructions on what one must do and must not do, as well as guidance on what language to avoid (on pain of committing blasphemy), what ideas to avoid (on pain of committing heresy), what symbols to avoid (on pain of committing idolatry). Necessarily but perhaps unfortunately, the Bible also explained how the world came into being in such literal detail that it could not accommodate new information produced by the telescope and subsequent technologies. The trials of Galileo and, three hundred years later, of Scopes were therefore about the admissibility of certain kinds of information. Both Cardinal Bellarmine and William Jennings Bryan were fighting to maintain the authority of the Bible to control information about the profane world as well as the sacred. In their defeat, more was lost than the Bible's claim to explain the origins and structure of nature. The Bible's authority in defining and categorizing moral behavior was also weakened.

Nonetheless, Scripture has at its core such a powerful mythology that even the residue of that mythology is still sufficient to serve as an exacting control mechanism for some people. It provides, first of all, a theory about the meaning of life and therefore rules on how one is to conduct oneself. With apologies to Rabbi Hillel, who expressed it more profoundly and in the time it takes to stand on one leg, the theory is as follows: There is one God, who created the universe and all that is in it. Although humans can never fully understand God, He has revealed Himself and His will to us throughout history, particularly through His commandments and the testament of the prophets as recorded in the Bible. The greatest of these commandments tells us that humans are to love God and express their love for Him through love, mercy, and justice to our fellow humans. At the end of time, all nations and humans will appear before God to be judged, and those who have followed His commandments will find favor in His sight. Those who have denied God and the commandments will perish utterly in the darkness that lies outside the presence of God's light.

To borrow from Hillel: That is the theory. All the rest is commentary.

Those who believe in this theory—particularly those who accept the Bible as the literal word of God—are free to dismiss other theories about the origin and meaning of life and to give minimal weight to the facts on which other theories are based. Moreover, in observing God's laws, and the detailed requirements of their enactment, believers receive guidance about what books they should not read, about what plays and films they should not see, about what music they should not hear, about what subjects their children should not study, and so on. For strict fundamentalists of the Bible, the theory and what follows from it seal them off from unwanted information, and in that way their actions are invested with meaning, clarity, and, they believe, moral authority.

Those who reject the Bible's theory and who believe, let us say, in the theory of Science are also protected from unwanted information. Their theory, for example, instructs them to disregard information about astrology, dianetics, and creationism, which they usually label as medieval superstition or subjective opinion. Their theory fails to give any guidance about moral information and, by definition, gives little weight to information that falls outside the constraints of science. Undeniably, fewer and fewer people are bound in any serious way to Biblical or other religious traditions as a source of compelling attention and authority, the result of which is that they make no moral decisions, only practical ones. This is still another way of defining Technopoly. The term is aptly used for a culture whose available theories do not offer guidance about what is acceptable information in the moral domain.

I trust the reader does not conclude that I am making an

argument for fundamentalism of any kind. One can hardly approve, for example, of a Muslim fundamentalism that decrees a death sentence to someone who writes what are construed as blasphemous words, or a Christian fundamentalism that once did the same or could lead to the same. I must hasten to acknowledge, in this context, that it is entirely possible to live as a Muslim, a Christian, or a Jew with a modified and temperate view of religious theory. Here, I am merely making the point that religious tradition serves as a mechanism for the regulation and valuation of information. When religion loses much or all of its binding power—if it is reduced to mere rhetorical ash—then confusion inevitably follows about what to attend to and how to assign it significance.

Indeed, as I write, another great world narrative, Marxism, is in the process of decomposing. No doubt there are fundamentalist Marxists who will not let go of Marx's theory, and will continue to be guided by its prescriptions and constraints. The theory, after all, is sufficiently powerful to have engaged the imagination and devotion of more than a billion people. Like the Bible, the theory includes a transcendent idea, as do all great world narratives. With apologies to a century and a half of philosophical and sociological disputation, the idea is as follows: All forms of institutional misery and oppression are a result of class conflict, since the consciousness of all people is formed by their material situation. God has no interest in this, because there is no God. But there *is* a plan, which is both knowable and beneficent. The plan unfolds in the movement of history itself, which shows unmistakably that the working class, in the end, must triumph. When it does, with or without the help of revolutionary movements, class itself will have disappeared. All will share equally in the bounties of nature and creative production, and no one will exploit the labors of another.

It is generally believed that this theory has fallen into disrepute among believers because information made available by television, films, telephone, fax machines, and other technologies has revealed that the working classes of capitalist nations are sharing quite nicely in the bounties of nature while at the same time enjoying a considerable measure of personal freedom. Their situation is so vastly superior to those of nations enacting Marxist theory that millions of people have concluded, seemingly all at once, that history may have no opinion whatever on the fate of the working class or, if it has, that it is moving toward a final chapter quite different in its point from what Marx prophesied.

All of this is said provisionally. History takes a long time, and there may yet be developments that will provide Marx's vision with fresh sources of verisimilitude. Meanwhile, the following points need to be made: Believers in the Marxist story were given quite clear guidelines on how they were to weight information and therefore to understand events. To the extent that they now reject the theory, they are threatened with conceptual confusion, which means they no longer know who to believe or what to believe. In the West, and especially in the United States, there is much rejoicing over this situation, and assurances are given that Marxism can be replaced by what is called "liberal democracy." But this must be stated more as a question than an answer, for it is no longer entirely clear what sort of story liberal democracy tells.

A clear and scholarly celebration of liberal democracy's triumph is found in Francis Fukuyama's essay "The End of History?" Using a somewhat peculiar definition of history, Fukuyama concludes that there will be no more ideological conflicts, all the competitors to modern liberalism having been defeated. In support of his conclusion, Fukuyama cites Hegel as having come to a similar position in the early nineteenth century, when the principles of liberty and equality, as expressed in the American and French revolutions, emerged triumphant. With the contemporary decline of fascism and communism, no threat now remains. But Fukuyama pays insufficient attention to the changes in meaning of liberal democracy over two centuries. Its meaning in a technocracy is quite different from its meaning in Technopoly; indeed, in Technopoly it comes much closer to what Walter Benjamin called "commodity capitalism." In the case of the United States, the great eighteenth-century revolution was not indifferent to commodity capitalism but was nonetheless infused with profound moral content. The United States was not merely an experiment in a new form of governance; it was the fulfillment of God's plan. True, Adams, Jefferson, and Paine rejected the supernatural elements in the Bible, but they never doubted that their experiment had the imprimatur of Providence. People were to be free but for a purpose. Their God-given rights implied obligations and responsibilities, not only to God but to other nations, to which the new republic would be a guide and a showcase of what is possible when reason and spirituality commingle.

It is an open question whether or not "liberal democracy" in its present form can provide a thought-world of sufficient moral substance to sustain meaningful lives. This is precisely the question that Václav Havel, then newly elected as president of Czechoslovakia, posed in an address to the U.S. Congress. "We still don't know how to put morality ahead of politics, science, and economics," he said. "We are still incapable of understanding that the only genuine backbone of our actions—if they are to be moral—is responsibility. Responsibility to something higher than my family, my country, my firm, my success." What Havel is saying is that it is not enough for his nation to liberate itself from one flawed theory; it is necessary to find another, and he worries that Technopoly provides no answer. To say it in still another way: Francis Fukuyama is wrong. There *is* another ideological conflict to be fought—between "liberal democracy" as conceived in the eighteenth century, with all its transcendent moral underpinnings, and Technopoly, a twentieth-century thought-world that functions not only without a transcendent narrative to provide moral underpinnings but also without strong social institutions to control the flood of information produced by technology.

Because that flood has laid waste the theories on which schools, families, political parties, religion, nationhood itself are based, American Technopoly must rely, to an obsessive extent, on technical methods to control the flow of information. Three such means merit special attention. They are interrelated but for purposes of clarity may be described separately.

The first is bureaucracy, which James Beniger in *The Control Revolution* ranks as "foremost among all technological solutions to the crisis of control."[4] Bureaucracy is not, of course, a creation of Technopoly. Its history goes back five thousand years, although the word itself did not appear in English until the nineteenth century. It is not unlikely that the ancient Egyptians found bureaucracy an irritation, but it is certain that, beginning in the nineteenth century, as bureaucracies became more important, the complaints against them became more insistent. John Stuart Mill referred to them as "administrative tyranny." Carlyle called them "the Continental nuisance." In a chilling paragraph, Tocqueville warned about them taking hold in the United States:

> I have previously made the distinction between two types of centralization, calling one governmental and the other administrative. Only the first exists in America, the second being almost unknown. If the directing power in American society had both these means of government at its disposal and combined the right to command with the faculty and habit to perform everything itself, if having established the general principles of the government, it entered into the details of their application, and having regulated the great interests of the country, it came down

> to consider even individual interest, then freedom would soon be banished from the New World.[5]

Writing in our own time, C. S. Lewis believed bureaucracy to be the technical embodiment of the Devil himself:

> I live in the Managerial Age, in a world of "Admin." The greatest evil is not now done in those sordid "dens of crime" that Dickens loved to paint. It is not done even in concentration camps and labour camps. In those we see its final result. But it is conceived and ordered (moved, seconded, carried, and minuted) in clean, carpeted, warmed, and well-lighted offices, by quiet men with white collars and cut fingernails and smooth-shaven cheeks who do not need to raise their voices. Hence, naturally enough, my symbol for Hell is something like the bureaucracy of a police state or the office of a thoroughly nasty business concern.[6]

Putting these attacks aside for the moment, we may say that in principle a bureaucracy is simply a coordinated series of techniques for reducing the amount of information that requires processing. Beniger notes, for example, that the invention of the standardized form—a staple of bureaucracy—allows for the "destruction" of every nuance and detail of a situation. By requiring us to check boxes and fill in blanks, the standardized form admits only a limited range of formal, objective, and impersonal information, which in some cases is precisely what is needed to solve a particular problem. Bureaucracy is, as Max Weber described it, an attempt to rationalize the flow of information, to make its use efficient to the highest degree by eliminating information that diverts attention from the problem at hand. Beniger offers as a prime example of such bureaucratic rationalization the decision in 1884 to organize time, on a worldwide basis, into twenty-four time zones. Prior to this decision, towns only a mile or two apart could and did differ on what time of day it was, which made the operation of railroads and other businesses unnecessarily complex. By simply ignoring the fact that solar time differs at each node of a transportation system, bureaucracy eliminated a problem of information chaos, much to the satisfaction of most people. But not of everyone. It must be noted that the idea of "God's own time" (a phrase used by the novelist Marie Corelli in the early twentieth century to oppose the introduction of Summer Time) had to be considered irrelevant. This is important to say, because, in attempting to make the most rational use of information, bureaucracy ignores all information and ideas that do not contribute to efficiency. The idea of God's time made no such contribution.

Bureaucracy is not in principle a social institution; nor are all institutions that reduce information by excluding some kinds or sources necessarily bureaucracies. Schools may exclude dianetics and astrology; courts exclude hearsay evidence. They do so for substantive reasons having to do with the theories on which these institutions are based. But bureaucracy has no intellectual, political, or moral theory—except for its implicit assumption that efficiency is the principal aim of all social institutions and that other goals are essentially less worthy, if not irrelevant. That is why John Stuart Mill thought bureaucracy a "tyranny" and C. S. Lewis identified it with Hell.

The transformation of bureaucracy from a set of techniques designed to serve social institutions to an autonomous meta-institution that largely serves itself came as a result of several developments in the mid- and late-nineteenth century: rapid industrial growth, improvements in transportation and communication, the extension of government into ever-larger realms of public and business affairs, the increasing centralization of governmental structures. To these were added, in the twentieth century, the information explosion and what we might call the "bureaucracy effect": as techniques for managing information became more necessary, extensive, and complex, the number of people and structures required to manage those techniques grew, and so did the amount of information *generated* by bureaucratic techniques. This created the need for bureaucracies to manage and coordinate bureaucracies, then for additional structures and techniques to manage the bureaucracies that coordinated bureaucracies, and so on—until bureaucracy became, to borrow again Karl Kraus's comment on psychoanalysis, the disease for which it purported to be the cure. Along the way, it ceased to be merely a servant of social institutions and became their master. Bureaucracy now not only solves problems but creates them. More important, it defines what our problems are—and they are always, in the bureaucratic view, problems of efficiency. As Lewis suggests, this makes bureaucracies exceedingly dangerous, because, though they were originally designed to process only technical information, they now are commonly employed to address problems of a moral, social, and political nature. The bureaucracy of the nineteenth century was largely concerned with making transportation, industry, and the distribution of goods more efficient. Technopoly's bureaucracy has broken loose from such restrictions and now claims sovereignty over all of society's affairs.

The peril we face in trusting social, moral, and political affairs to bureaucracy may be highlighted by reminding ourselves what a bureaucrat does. As the word's history suggests, a bureaucrat is little else than a glorified counter. The French word *bureau* first meant a cloth for covering a reckoning table, then the table itself, then the room in which the table was kept, and finally the office and staff that ran the entire counting room or house. The word "bureaucrat" has come to mean a person who by training, commitment, and even temperament is indifferent to both the content and the totality of a human problem. The bureaucrat considers the implications of a decision only to the extent that the decision will affect the efficient operations of the bureaucracy, and takes no responsibility for its human consequences. Thus, Adolf Eichmann becomes the basic model and metaphor for a bureaucrat in the age of Technopoly.[7] When faced with the charge of crimes against humanity, he argued that he had no part in the formulation of Nazi political or sociological theory; he dealt only with the technical problems of moving vast numbers of people from one place to another. Why they were being moved and, especially, what would happen to them when they arrived at their destination were not relevant to his job. Although the jobs of bureaucrats in today's Technopoly have results far less horrific, Eichmann's answer is probably given five thousand times a day in America alone: I have no responsibility for the human consequences of my decisions. I am only responsible for the efficiency of my part of the bureaucracy, which must be maintained at all costs.

Eichmann, it must also be noted, was an expert. And expertise is a second important technical means by which Technopoly strives furiously to control information. There have, of course, always been experts, even in tool-using cultures. The pyramids, Roman roads, the Strasbourg Cathedral, could hardly have been built without experts. But the expert in Technopoly has two characteristics that distinguish him or her from experts of the past. First, Technopoly's experts tend to be ignorant about any matter not directly related to their specialized area. The average psychotherapist, for example, barely has even superficial knowledge of literature, philosophy, social history, art, religion, and biology, and is not expected to have such knowledge. Second, like bureaucracy itself (with which an expert may or may not be connected), Technopoly's experts claim dominion not only over technical matters but also over social, psychological, and moral affairs. In the United States, we have experts in how to raise children, how to educate them, how to be lovable, how to make love, how to influence people, how to make friends. There is no aspect of human relations that has not been technicalized and therefore relegated to the control of experts.

These special characteristics of the expert arose as a result of three factors. First, the growth of bureaucracies, which, in effect, produced the world's first entirely mechanistic specialists and thereby gave credence and prestige to the specialist-as-ignoramus. Second, the weakening of traditional social institutions, which led ordinary people to lose confidence in the value of tradition. Third, and underlying everything else, the torrent of information which made it impossible for anyone to possess more than a tiny fraction of the sum total of human knowledge. As a college undergraduate, I was told by an enthusiastic professor of German literature that Goethe was the last person who knew everything. I assume she meant, by this astounding remark, less to deify Goethe than to suggest that by the year of his death, 1832, it was no longer possible for even the most brilliant mind to comprehend, let alone integrate, what was known.

The role of the expert is to concentrate on one field of knowledge, sift through all that is available, eliminate that which has no bearing on a problem, and use what is left to assist in solving a problem. This process works fairly well in situations where only a technical solution is required and there is no conflict with human purposes—for example, in space rocketry or the construction of a sewer system. It works less well in situations where technical requirements may conflict with human purposes, as in medicine or architecture. And it is disastrous when applied to situations that cannot be solved by technical means and where efficiency is usually irrelevant, such as in education, law, family life, and problems of personal maladjustment. I assume I do not need to convince the reader that there are no experts—there can be no experts—in child-rearing and lovemaking and friend-making. All of this is a figment of the Technopolist's imagination, made plausible by the use of technical machinery, without which the expert would be totally disarmed and exposed as an intruder and an ignoramus.

Technical machinery is essential to both the bureaucrat and the expert, and may be regarded as a third mechanism of information control. I do not have in mind such "hard" technologies as the computer—which must, in any case, be treated separately, since it embodies all that Technopoly stands for. I have in mind "softer" technologies such as IQ tests, SATs, standardized forms, taxonomies, and opinion polls. Some of these I discuss in detail in chapter eight, "Invisible Technologies," but I mention them here because their role in reducing the types and quantity of information admitted to a system often goes unnoticed, and therefore their role in redefining traditional concepts also goes unnoticed. *There is, for example, no test that can measure a person's intelligence.* Intelligence is a general term used to denote one's capacity to solve real-life problems in a variety of novel contexts. It is acknowledged by everyone except experts that each person varies greatly in such capacities, from consistently effective to consistently ineffective, depending on the kinds of problems requiring solution. If, however, we are made to believe that a test can reveal precisely the quantity of intelligence a person has, then, for all institutional purposes, a score on a test becomes his or her intelligence. The test transforms an abstract and multifaceted meaning into a technical and exact term that leaves out everything of importance. One might even say that an intelligence test is a tale told by an expert, signifying nothing. Nonetheless, the expert relies on our believing in the reality of technical machinery, which means we will reify the answers generated by the machinery. We come to believe that our score *is* our intelligence, or our capacity for creativity or love or pain. We come to believe that the results of opinion polls *are* what people believe, as if our beliefs can be encapsulated in such sentences as "I approve" and "I disapprove."

When Catholic priests use wine, wafers, and incantations to embody spiritual ideas, they acknowledge the mystery and the metaphor being used. But experts of Technopoly acknowledge no such overtones or nuances when they use forms, standardized tests, polls, and other machinery to give technical reality to ideas about intelligence, creativity, sensitivity, emotional imbalance, social deviance, or political opinion. They would have us believe that technology can plainly reveal the true nature of some human condition or belief because the score, statistic, or taxonomy has given it technical form.

There is no denying that the technicalization of terms and problems is a serious form of information control. Institutions can make decisions on the basis of scores and statistics, and there certainly may be occasions where there is no reasonable alternative. But unless such decisions are made with profound skepticism—that is, acknowledged as being made for administrative convenience—they are delusionary. In Technopoly, the delusion is sanctified by our granting inordinate prestige to experts who are armed with sophisticated technical machinery. Shaw once remarked that all professions are conspiracies against the laity. I would go further: in Technopoly, all experts are invested with the charisma of priestliness. Some of our priest-experts are called psychiatrists, some psychologists, some sociologists, some statisticians. The god they serve does not speak of righteousness or goodness or mercy or grace. Their god speaks of efficiency, precision, objectivity. And that is why such concepts as sin and evil disappear in Technopoly. They come from a moral universe that is irrelevant to the theology of expertise. And so the priests of Technopoly call sin "social deviance," which is a statistical concept, and they call evil "psychopathology," which is a medical concept. Sin and evil disappear because they cannot be measured and objectified, and therefore cannot be dealt with by experts.

As the power of traditional social institutions to organize

perceptions and judgment declines, bureaucracies, expertise, and technical machinery become the principal means by which Technopoly hopes to control information and thereby provide itself with intelligibility and order. The rest of this book tells the story of why this cannot work, and of the pain and stupidity that are the consequences.

6

The Ideology of Machines: Medical Technology

A few years ago, an enterprising company made available a machine called HAGOTH, of which it might be said, this was Technopoly's most ambitious hour. The machine cost $1,500, the bargain of the century, for it was able to reveal to its owner whether someone talking on the telephone was telling the truth. It did this by measuring the "stress content" of a human voice as indicated by its oscillations. You connected HAGOTH to your telephone and, in the course of conversation, asked your caller some key question, such as "Where did you go last Saturday night?" HAGOTH had sixteen lights—eight green and eight red—and when the caller replied, HAGOTH went to work. Red lights went on when there was much stress in the voice, green lights when there was little. As an advertisement for HAGOTH said, "Green indicates no stress, hence truthfulness." In other words, according to HAGOTH, it is not possible to speak the truth in a quivering voice or to lie in a steady one—an idea that would doubtless amuse Richard Nixon. At the very least, we must say that HAGOTH's definition of truthfulness was peculiar, but so precise and exquisitely technical as to command any bureaucrat's admiration. The same may be said of the definition of intelligence as expressed in a standard-brand intelligence test. In fact, an intelligence test works exactly like HAGOTH. You connect a pencil to the fingers of a young person and address some key questions to him or her; from the replies a computer can calculate exactly how much intelligence exists in the young person's brain.[1]

HAGOTH has mercifully disappeared from the market, for what reason I do not know. Perhaps it was sexist or culturally biased or, worse, could not measure oscillations accurately enough. When it comes to machinery, what Technopoly insists upon most is accuracy. The idea embedded in the machine is largely ignored, no matter how peculiar.

Though HAGOTH has disappeared, its idea survives—for example, in the machines called "lie detectors." In America, these are taken very seriously by police officers, lawyers, and corporate executives who ever more frequently insist that their employees be subjected to lie-detector tests. As for intelligence tests, they not only survive but flourish, and have been supplemented by vocational aptitude tests, creativity tests, mental-health tests, sexual-attraction tests, and even marital-compatibility tests. One would think that two people who have lived together for a number of years would have noticed for themselves whether they get along or not. But in Technopoly, these subjective forms of knowledge have no official status, and must be confirmed by tests administered by experts. Individual judgments, after all, are notoriously unreliable, filled with ambiguity and plagued by doubt, as Frederick W. Taylor warned. Tests and machines are not. Philosophers may agonize over the questions "What is truth?" "What is intelligence?" "What is the good life?" But in Technopoly there is no need for such intellectual struggle. Machines eliminate complexity, doubt, and ambiguity. They work swiftly, they are standardized, and they provide us with numbers that you can see and calculate with.

They tell us that when eight green lights go on someone is speaking the truth. That is all there is to it. They tell us that a score of 136 means more brains than a score of 104. This is Technopoly's version of magic.

What is significant about magic is that it directs our attention to the wrong place. And by doing so, evokes in us a sense of wonder rather than understanding. In Technopoly, we are surrounded by the wondrous effects of machines and are encouraged to ignore the ideas embedded in them. Which means we become blind to the ideological meaning of our technologies. In this chapter and the next, I should like to provide examples of how technology directs us to construe the world.

In considering here the ideological biases of medical technology, let us begin with a few relevant facts. Although the U.S. and England have equivalent life-expectancy rates, American doctors perform six times as many cardiac bypass operations per capita as English doctors do. American doctors perform more diagnostic tests than doctors do in France, Germany, or England. An American woman has two to three times the chance of having a hysterectomy as her counterpart in Europe; 60 percent of the hysterectomies performed in America are done on women under the age of forty-four. American doctors do more prostate surgery per capita than do doctors anywhere in Europe, and the United States leads the industrialized world in the rate of cesarean-section operations—50 to 200 percent higher than in most other countries. When American doctors decide to forgo surgery in favor of treatment by drugs, they give higher dosages than doctors elsewhere. They prescribe about twice as many antibiotics as do doctors in the United Kingdom and commonly prescribe antibiotics when bacteria are likely to be present, whereas European doctors tend to prescribe antibiotics only if they know that the infection is caused by bacteria *and* is also serious.[2] American doctors use far more X-rays per patient than do doctors in other countries. In one review of the extent of X-ray use, a radiologist discovered cases in which fifty to one hundred X-rays had been taken of a single patient when five would have been sufficient. Other surveys have shown that, for almost one-third of the patients, the X-ray could have been omitted or deferred on the basis of available clinical data.[3]

The rest of this chapter could easily be filled with similar statistics and findings. Perhaps American medical practice is best summarized by the following warning, given by Dr. David E. Rogers in a presidential address to the Association of American Physicians:

> As our interventions have become more searching, they have also become more costly and more hazardous. Thus, today it is not unusual to find a fragile elder who walked into the hospital, [and became] slightly confused, dehydrated, and somewhat the worse for wear on the third hospital day because his first 48 hours in the hospital were spent undergoing a staggering series of exhausting diagnostic studies in various laboratories or in the radiology suite.[4]

None of this is surprising to anyone familiar with American medicine, which is notorious for its characteristic "aggressiveness." The question is, why? There are three interrelated reasons, all relevant to the imposition of machinery. The first has to do with the American character, which I have previously discussed as being so congenial to the sovereignty of technology. In *Medicine and Culture,* Lynn Payer describes it in the following way:

> The once seemingly limitless lands gave rise to a spirit that anything was possible if only the natural environment . . . could be conquered. Disease could also be conquered,

> but only by aggressively ferreting it out diagnostically and just as aggressively treating it, preferably by taking something out rather than adding something to increase the resistance.[5]

To add substance to this claim, Ms. Payer quotes Oliver Wendell Holmes as saying, with his customary sarcasm:

> How could a people which has a revolution once in four years, which has contrived the Bowie Knife and the revolver . . . which insists in sending out yachts and horses and boys to outsail, outrun, outfight and checkmate all the rest of creation; how could such a people be content with any but "heroic" practice? What wonder that the stars and stripes wave over doses of ninety grams of sulphate of quinine and that the American eagle screams with delight to see three drachms [180 grains] of calomel given at a single mouthful?[6]

The spirit of attack mocked here by Holmes was given impetus even before the American Revolution by Dr. Benjamin Rush, perhaps the most influential medical man of his age. Rush believed that medicine had been hindered by doctors placing "undue reliance upon the powers of nature in curing disease," and specifically blamed Hippocrates and his tradition for this lapse. Rush had considerable success in curing patients of yellow fever by prescribing large quantities of mercury and performing purges and bloodletting. (His success was probably due to the fact that the patients either had mild cases of yellow fever or didn't have it at all.) In any event, Rush was particularly enthusiastic about bleeding patients, perhaps because he believed that the body contained about twenty-five pints of blood, which is more than twice the average actual amount. He advised other doctors to continue bleeding a patient until four-fifths of the body's blood was removed. Although Rush was not in attendance during George Washington's final days, Washington was bled seven times on the night he died, which, no doubt, had something to do with why he died. All of this occurred, mind you, 153 years after Harvey discovered that blood circulates throughout the body.

Putting aside the question of the available medical knowledge of the day, Rush was a powerful advocate of action—indeed, gave additional evidence of his aggressive nature by being one of the signers of the Declaration of Independence. He persuaded both doctors and patients that American diseases were tougher than European diseases and required tougher treatment. "Desperate diseases require desperate remedies" was a phrase repeated many times in American medical journals in the nineteenth century. The Americans, who considered European methods to be mild and passive—one might even say effeminate—met the challenge by eagerly succumbing to the influence of Rush: they accepted the imperatives to intervene, to mistrust nature, to use the most aggressive therapies available. The idea, as Ms. Payer suggests, was to conquer both a continent and the diseases its weather and poisonous flora and fauna inflicted.

So, from the outset, American medicine was attracted to new technologies. Far from being "neutral," technology was to be the weapon with which disease and illness would be vanquished. The weapons were not long in coming. The most significant of the early medical technologies was the stethoscope, invented (one might almost say discovered) by the French physician René-Théophile-Hyacinthe Laënnec in 1816. The circumstances surrounding the invention are worth mentioning.

Working at the Necker Hospital in Paris, Laënnec was examining a young woman with a puzzling heart disorder. He tried to use percussion and palpation (pressing the hand upon the body in hope of detecting internal abnormalities), but the patient's obesity made this ineffective. He next considered auscultation (placing his ear on the patient's chest to hear the heart beat), but the patient's youth and sex discouraged him. Laënnec then remembered that sound traveling through solid bodies is amplified. He rolled some sheets of paper into a cylinder, placed one end on the patient's chest and the other to his ear. *Voilà!* The sounds he heard were clear and distinct. "From this moment," he later wrote, "I imagined that the circumstance might furnish means for enabling us to ascertain the character, not only of the action of the heart, but of every species of sound produced by the motion of all the thoracic viscera." Laënnec worked to improve the instrument, eventually using a rounded piece of wood, and called it a "stethoscope," from the Greek words for "chest" and "I view."[7]

For all its simplicity, Laënnec's invention proved extraordinarily useful, particularly in the accuracy with which it helped to diagnose lung diseases like tuberculosis. Chest diseases of many kinds were no longer concealed: the physician with a stethoscope could, as it were, conduct an autopsy on the patient while the patient was still alive.

But it should not be supposed that all doctors or patients were enthusiastic about the instrument. Patients were often frightened at the sight of a stethoscope, assuming that its presence implied imminent surgery, since, at the time, only surgeons used instruments, not physicians. Doctors had several objections, ranging from the trivial to the significant. Among the trivial was the inconvenience of carrying the stethoscope, a problem some doctors solved by carrying it, crosswise, inside their top hats. This was not without its occasional embarrassments—an Edinburgh medical student was accused of possessing a dangerous weapon when his stethoscope fell out of his hat during a snowball fight. A somewhat less trivial objection raised by doctors was that if they used an instrument they would be mistaken for surgeons, who were then considered mere craftsmen. The distinction between physicians and surgeons was unmistakable then, and entirely favorable to physicians, whose intellect, knowledge, and insight were profoundly admired. It is perhaps to be expected that Oliver Wendell Holmes, professor of anatomy at Harvard and always a skeptic about aggressiveness in medicine, raised objections about the overzealous use of the stethoscope; he did so, in characteristic fashion, by writing a comic ballad, "The Stethoscope Song," in which a physician makes several false diagnoses because insects have nested in his stethoscope.

But a serious objection raised by physicians, and one which has resonated throughout the centuries of technological development in medicine, is that interposing an instrument between patient and doctor would transform the practice of medicine; the traditional methods of questioning patients, taking their reports seriously, and making careful observations of exterior symptoms would become increasingly irrelevant. Doctors would lose their ability to conduct skillful examinations and rely more on machinery than on their own experience and insight. In his detailed book *Medicine and the Reign of Technology,* Stanley Joel Reiser compares the effects of the stethoscope to the effects of the printing press on Western culture. The printed book, he argues, helped to create the detached and objective thinker. Similarly, the stethoscope

> helped to create the objective physician, who could move away from involvement with the patient's experiences and sensations, to a more detached relation, less with the patient but more with the sounds from within the body. Undistracted by the motives and beliefs of the patient, the auscultator [another term for the stethoscope] could make

> a diagnosis from sounds that he alone heard emanating from body organs, sounds that he believed to be objective, bias-free representations of the disease process.[8]

Here we have expressed two of the key *ideas* promoted by the stethoscope: Medicine is about disease, not the patient. And, what the patient knows is untrustworthy; what the machine knows is reliable.

The stethoscope could not by itself have made such ideas stick, especially because of the resistance to them, even in America, by doctors whose training and relationship to their patients led them to oppose mechanical interpositions. But the ideas were amplified with each new instrument added to the doctor's arsenal: the ophthalmoscope (invented by Hermann von Helmholtz in 1850), which allowed doctors to see into the eye; the laryngoscope (designed by Johann Czermak, a Polish professor of physiology, in 1857), which allowed doctors to inspect the larynx and other parts of the throat, as well as the nose; and, of course, the X-ray (developed by Wilhelm Roentgen in 1895), which could penetrate most substances but not bones. "If the hand be held before the fluorescent screen," Roentgen wrote, "the shadow shows the bones darkly with only faint outlines of the surrounding tissues." Roentgen was able to reproduce this effect on photographic plates and make the first X-ray of a human being, his wife's hand.

By the turn of the century, medicine was well on its way to almost total reliance on technology, especially after the development of diagnostic laboratories and the discovery and use of antibiotics in the 1940s. Medical practice had entered a new stage. The first had been characterized by direct communication with the patient's experiences based on the patient's reports, and the doctor's questions and observations. The second was characterized by direct communication with patients' bodies through physical examination, including the use of carefully selected technologies. The stage we are now in is characterized by indirect communication with the patient's experience and body through technical machinery. In this stage, we see the emergence of specialists—for example, pathologists and radiologists—who interpret the meaning of technical information and have no connection whatsoever with the patient, only with tissue and photographs. It is to be expected that, as medical practice moved from one stage to another, doctors tended to lose the skills and insights that predominated in the previous stage. Reiser sums up what this means:

> So, without realizing what has happened, the physician in the last two centuries has gradually relinquished his unsatisfactory attachment to subjective evidence—what the patient says—only to substitute a devotion to technological evidence—what the machine says. He has thus exchanged one partial view of disease for another. As the physician makes greater use of the technology of diagnosis, he perceives his patient more and more indirectly through a screen of machines and specialists; he also relinquishes control over more and more of the diagnostic process. These circumstances tend to estrange him from his patient and from his own judgment.[9]

There is still another reason why the modern physician is estranged from his own judgment. To put it in the words of a doctor who remains skilled in examining his patients and in evaluating their histories: "Everyone who has a headache wants and expects a CAT scan." He went on to say that roughly six out of every ten CAT scans he orders are unnecessary, with no basis in the clinical evidence and the patient's reported experience and sensations. Why are they done? As a protection against malpractice suits. Which is to say, as medical practice has moved into the stage of total reliance on machine-generated information, so have the patients. Put simply, if a patient does not obtain relief from a doctor who has failed to use all the available technological resources, including drugs, the doctor is deemed vulnerable to the charge of incompetence. The situation is compounded by the fact that the personal relationship between doctor and patient now, in contrast to a century ago, has become so arid that the patient is not restrained by intimacy or empathy from appealing to the courts. Moreover, doctors are reimbursed by medical-insurance agencies on the basis of what they *do*, not on the amount of time they spend with patients. Nontechnological medicine is time-consuming. It is more profitable to do a CAT scan on a patient with a headache than to spend time getting information about his or her experiences and sensations.

What all this means is that even restrained and selective technological medicine becomes very difficult to do, economically undesirable, and possibly professionally catastrophic. The culture itself—its courts, its bureaucracies, its insurance system, the training of doctors, patients' expectations—is organized to support technological treatments. There are no longer methods of treating illness; there is only one method—the technological one. Medical competence is now defined by the quantity and variety of machinery brought to bear on disease.

As I remarked, three interrelated reasons converged to create this situation. The American character was biased toward an aggressive approach and was well prepared to accommodate medical technology; the nineteenth-century technocracies, obsessed with invention and imbued with the idea of progress, initiated a series of remarkable and wondrous inventions; and the culture reoriented itself to ensure that technological aggressiveness became the basis of medical practice. The ideas promoted by this domination of technology can be summed up as follows: Nature is an implacable enemy that can be subdued only by technical means; the problems created by technological solutions (doctors call these "side effects") can be solved only by the further application of technology (we all know the joke about an amazing new drug that cures nothing but has interesting side effects); medical practice must focus on disease, not on the patient (which is why it is possible to say that the operation or therapy was successful but the patient died); and information coming from the patient cannot be taken as seriously as information coming from a machine, from which it follows that a doctor's judgment, based on insight and experience, is less worthwhile than the calculations of his machinery.

Do these ideas lead to better medicine? In some respects, yes; in some respects, no. The answer tends to be "yes" when one considers how doctors now use lasers to remove cataracts quickly, painlessly, and safely; or how they can remove a gallbladder by using a small television camera (a laparoscope) inserted through an equally small puncture in the abdomen to guide the surgeon's instruments to the diseased organ through still another small puncture, thus making it unnecessary to cut open the abdomen. Of course, those who are inclined to answer "no" to the question will ask how many laparoscopic cholecystectomies are performed *because* of the existence of the technology. This is a crucial point.

Consider the case of cesarean sections. Close to one out of every four Americans is now born by C-section. Through modern technology, American doctors can deliver babies who would have died otherwise. As Dr. Laurence Horowitz notes in *Taking Charge of Your Medical Fate*, ". . . the proper goal of C-sections is to improve the chances of babies at risk, and that goal has been achieved."[10] But C-sections are a surgical procedure, and when they are done routinely as an elective option, there is considerable and unnecessary danger; the chances of a woman's dying during a C-section delivery are two to four times greater than during a normal vaginal delivery. In other words, C-sections can and do save the lives of babies at risk, but when they are done for other reasons—for example, for the convenience of doctor or mother—they pose an unnecessary threat to health, and even life.

To take another example: a surgical procedure known as carotid endarterectomy is used to clean out clogged arteries, thus reducing the likelihood of stroke. In 1987, more than one hundred thousand Americans had this operation. It is now established that the risks involved in such surgery outweigh the risks of suffering a stroke. Horowitz again: "In other words, for certain categories of patients, the operation may actually kill more people than it saves."[11] To take still another example: about seventy-eight thousand people every year get cancer from medical and dental X-rays. In a single generation, it is estimated, radiation will induce 2.34 million cancers.[12]

Examples of this kind can be given with appalling ease. But in the interests of fairness the question about the value of technology in medicine is better phrased in the following way: Would American medicine be better were it not so totally reliant on the technological imperative? Here the answer is clearly, yes. We know, for example, from a Harvard Medical School study which focused on the year 1984 (no Orwellian reference intended), that in New York State alone there were thirty-six thousand cases of medical negligence, including seven thousand deaths related in some way to negligence. Although the study does not give figures on what kinds of negligence were found, the example is provided of doctors prescribing penicillin without asking the patients whether they were hypersensitive to the drug. We can assume that many of the deaths resulted not only from careless prescriptions and the doctors' ignorance of their patients' histories but also from unnecessary surgery. In other words, iatrogenics (treatment-induced illness) is now a major concern for the profession, and an even greater concern for the patient. Doctors themselves feel restricted and dominated by the requirement to use all available technology.

And patients may be justifiably worried by reports that quite possibly close to 40 percent of the operations performed in America are not necessary. In *Health Shock*, Martin Weitz cites the calculations of Professor John McKinlay that more deaths are caused by surgery each year in the United States than the annual number of deaths during the wars in Korea and Vietnam. As early as 1974, a Senate investigation into unnecessary surgery reported that American doctors had performed 2.4 million unnecessary operations, causing 11,900 deaths and costing about $3.9 billion.[13] We also know that, in spite of advanced technology (quite possibly because of it), the infant-survival rate in the United States ranks only fourteenth in the world, and it is no exaggeration to say that American hospitals are commonly regarded as among the most dangerous places in the nation. It is also well documented that, wherever doctor strikes have occurred, the mortality rate declines.

There are, one may be sure, very few doctors who are satisfied with technology's stranglehold on medical practice. And there are far too many patients who have been its serious victims. What conclusions may we draw? First, technology is not a neutral element in the practice of medicine: doctors do not merely use technologies but are used by them. Second, technology creates its own imperatives and, at the same time, creates a wide-ranging social system to reinforce its imperatives. And third, technology changes the practice of medicine by redefining what doctors are, redirecting where they focus their attention, and reconceptualizing how they view their patients and illness.

Like some well-known diseases, the problems that have arisen as a result of the reign of technology came slowly and were barely perceptible at the start. As technology grew, so did the influence of drug companies and the manufacturers of medical instruments. As the training of doctors changed, so did the expectations of patients. As the increase in surgical procedures multiplied, so did the diagnoses which made them seem necessary. Through it all, the question of what was being *undone* had a low priority if it was asked at all. The Zeitgeist of the age placed such a question in a range somewhere between peevishness and irrelevance. In a growing Technopoly, there is no time or inclination to speak of technological debits.

7

The Ideology of Machines: Computer Technology

That American Technopoly has now embraced the computer in the same hurried and mindless way it embraced medical technology is undeniable, was perhaps inevitable, and is certainly most unfortunate. This is not to say that the computer is a blight on the symbolic landscape; only that, like medical technology, it has usurped powers and enforced mind-sets that a fully attentive culture might have wished to deny it. Thus, an examination of the ideas embedded in computer technology is worth attempting. Others, of course, have done this, especially Joseph Weizenbaum in his great and indispensable book *Computer Power and Human Reason.* Weizenbaum, however, ran into some difficulties, as everyone else has, because of the "universality" of computers, meaning (a) that their uses are infinitely various, and (b) that computers are commonly integrated into the structure of other machines. It is, therefore, hard to isolate specific ideas promoted by computer technology. The computer, for example, is quite unlike the stethoscope, which has a limited function in a limited context. Except for safecrackers, who, I am told, use stethoscopes to hear the tumblers of locks click into place, stethoscopes are used only by doctors. But everyone uses or is used by computers, and for purposes that seem to know no boundaries.

Putting aside such well-known functions as electronic filing, spreadsheets, and word-processing, one can make a fascinating list of the innovative, even bizarre, uses of computers. I have before me a report from *The New York Times* that tells us how computers are enabling aquatic designers to create giant water slides that mimic roller coasters and eight-foot-high artificial waves.[1] In my modest collection, I have another article about the uses of personal computers for making presentations at corporate board meetings.[2] Another tells of how computer graphics help jurors to remember testimony better. Gregory Mazares, president of the graphics unit of Litigation Sciences, is quoted as saying, "We're a switched-on, tuned-in, visually oriented society, and jurors tend to believe what they see. This technology keeps the jury's attention by simplifying the material and by giving them little bursts of information."[3] While Mr. Mazares is helping switched-on people to remember things, Morton David, chief executive officer of Franklin Computer, is helping them find any word in the Bible with lightning speed by producing electronic Bibles. (The word "lightning," by the way, appears forty-two times in the New International version and eight times in the King James version. Were you so inclined, you could discover this for yourself in a matter of seconds.) This fact so dominates Mr. David's imagination that he is quoted as saying, "Our technology may have made a change as momentous as the Gutenberg invention of movable type."[4] And then there is an article that reports a computer's use to make investment decisions, which helps you, among other things, to create "what-if" scenarios, although with how much accuracy we are not told.[5] In *Technology Review*, we find a description of how computers are used to help the police locate the addresses of callers in distress; a prophecy is made that in time police officers will have so much instantly available information about any caller that they will know how seriously to regard the caller's appeal for help.

One may well wonder if Charles Babbage had any of this in mind when he announced in 1822 (only six years after the appearance of Laënnec's stethoscope) that he had invented a machine capable of performing simple arithmetical calculations. Perhaps he did, for he never finished his invention and started work on a more ambitious machine, capable of doing more complex tasks. He abandoned that as well, and in 1833 put aside his calculator project completely in favor of a programmable machine that became the forerunner of the modern computer. His first such machine, which he characteristically never finished, was to be controlled by punch cards adapted from devices French weavers used to control thread sequences in their looms.

Babbage kept improving his programmable machine over the next thirty-seven years, each design being more complex than the last.[6] At some point, he realized that the mechanization of numerical operations gave him the means to manipulate non-numerical symbols. It is not farfetched to say that Babbage's insight was comparable to the discovery by the Greeks in the third century B.C. of the principle of alphabetization—that is, the realization that the symbols of the alphabet could be separated from their phonetic function and used as a system for the classification, storage, and retrieval of information. In any case, armed with his insight, Babbage was able to speculate about the possibility of designing "intelligent" information machinery, though the mechanical technology of his time was inadequate to allow the fulfillment of his ideas. The computer as we know it today had to await a variety of further discoveries and inventions, including the telegraph, the telephone, and the application of Boolean algebra to relay-based circuitry, resulting in Claude Shannon's creation of digital logic circuitry. Today, when the word "computer" is used without a modifier before it, it normally refers to some version of the machine invented by John von Neumann in the 1940s. Before that, the word "computer" referred to a person (similarly to the early use of the word "typewriter") who performed some kind of mechanical calculation. As calculation shifted from people to machines, so did the word, especially because of the power of von Neumann's machine.

Certainly, after the invention of the digital computer, it was abundantly clear that the computer was capable of performing functions that could in some sense be called "intelligent." In 1936, the great English mathematician Alan Turing showed that it was possible to build a machine that would, for many practical purposes, behave like a problem-solving human being. Turing claimed that he would call a machine "intelligent" if, through typed messages, it could exchange thoughts with a human being—that is, hold up its end of a conversation. In the early days of MIT's Artificial Intelligence Laboratory, Joseph Weizenbaum wrote a program called ELIZA, which showed how easy it was to meet Turing's test for intelligence. When asked a question with a proper noun in it, ELIZA's program could respond with "Why are you interested in," followed by the proper noun and a question mark. That is, it could invert statements and seek more information about one of the nouns in the statement. Thus, ELIZA acted much like a Rogerian psychologist, or at least a friendly and inexpensive therapist. Some people who used ELIZA refused to believe that they were conversing with a mere machine. Having, in effect, created a Turing machine, Weizenbaum eventually pulled the program off the computer network and was stimulated to write *Computer Power and Human Reason*, in which, among other things, he raised questions about the research programs of those working in artificial intelligence; the assumption that whatever a computer *can* do, it *should* do; and the effects of computer technology on the way people construe the world—that is, the ideology of the computer, to which I now turn.

The most comprehensive idea conveyed by the computer is suggested by the title of J. David Bolter's book, *Turing's Man.* His title is a metaphor, of course, similar to what would be suggested by saying that from the sixteenth century until recently we were "Gutenberg's Men." Although Bolter's main practical interest in the computer is in its function as a new kind of book, he argues that it is the dominant metaphor of our age; it defines our age by suggesting a new relationship to information, to work, to power, and to nature itself. That relationship can best be described by saying that the computer redefines humans as "information processors" and nature itself as information to be processed. The fundamental metaphorical message of the computer, in short, is that we are machines—thinking machines, to be sure, but machines nonetheless. It is for this reason that the computer is the quintessential, incomparable, near-perfect machine for Technopoly. It subordinates the claims of our nature, our biology, our emotions, our spirituality. The computer claims sovereignty over the whole range of human experience, and supports its claim by showing that it "thinks" better than we can. Indeed, in his almost hysterical enthusiasm for artificial intelligence, Marvin Minsky has been quoted as saying that the thinking power of silicon "brains" will be so formidable that "If we are lucky, they will keep us as pets."[7] An even giddier remark, although more dangerous, was offered by John McCarthy, the inventor of the term "artificial intelligence." McCarthy claims that "even machines as simple as thermostats can be said to have beliefs." To the obvious question, posed by the philosopher John Searle, "What beliefs does your thermostat have?," McCarthy replied, "My thermostat has three beliefs—it's too hot in here, it's too cold in here, and it's just right in here."[8]

What is significant about this response is that it has redefined the meaning of the word "belief." The remark rejects the view that humans have internal states of mind that are the foundation of belief and argues instead that "belief" means only what someone or something does. The remark also implies that simulating an idea is synonymous with duplicating the idea. And, most important, the remark rejects the idea that mind is a biological phenomenon.

In other words, what we have here is a case of metaphor gone mad. From the proposition that humans are in some respects like machines, we move to the proposition that humans are little else but machines and, finally, that human beings *are* machines. And then, inevitably, as McCarthy's remark suggests, to the proposition that machines are human beings. It follows that machines can be made that duplicate human intelligence, and thus research in the field known as artificial intelligence was inevitable. What is most significant about this line of thinking is the dangerous reductionism it represents. Human intelligence, as Weizenbaum has tried energetically to remind everyone, is not transferable. The plain fact is that humans have a unique, biologically rooted, intangible mental life which in some limited respects can be simulated by a machine but can never be duplicated. Machines cannot feel and, just as important, cannot *understand.* ELIZA can ask, "Why are you worried about your mother?," which might be exactly the question a therapist would ask. But the machine does not know what the question means or even *that* the question means. (Of course, there may be some therapists who do not know what the question means either, who ask it routinely, ritualistically, inattentively. In that case we may say they are acting like a machine.) It is meaning, not utterance, that makes mind unique. I use "meaning" here to refer to something more than the result of putting together symbols the denotations of which are commonly shared by at least two people. As I understand it, meaning also includes those things we call feelings, experiences, and sensations that do not have to be, and sometimes cannot be, put into symbols. They "mean" nonetheless. Without concrete symbols, a computer is merely a pile of junk. Although the quest for a machine that duplicates mind has ancient roots, and although digital logic circuitry has given that quest a scientific structure, artificial intelligence does not and cannot lead to a meaning-making, understanding, and feeling creature, which is what a human being is.

All of this may seem obvious enough, but the metaphor of the machine as human (or the human as machine) is sufficiently powerful to have made serious inroads in everyday language. People now commonly speak of "programming" or "deprogramming" themselves. They speak of their brains as a piece of "hard wiring," capable of "retrieving data," and it has become common to think about thinking as a mere matter of processing and decoding.

Perhaps the most chilling case of how deeply our language is absorbing the "machine as human" metaphor began on November 4, 1988, when the computers around the ARPANET network became sluggish, filled with extraneous data, and then clogged completely. The problem spread fairly quickly to six thousand computers across the United States and overseas. The early hypothesis was that a software program had attached itself to other programs, a situation which is called (in another human-machine metaphor) a "virus." As it happened, the intruder was a self-contained program explicitly designed to disable computers, which is called a "worm." But the technically incorrect term "virus" stuck, no doubt because of its familiarity and its human connections. As Raymond Gozzi, Jr., discovered in his analysis of how the mass media described the event, newspapers noted that the computers were "infected," that the virus was "virulent" and "contagious," that attempts were made to "quarantine" the infected computers, that attempts were also being made to "sterilize" the network, and that programmers hoped to develop a "vaccine" so that computers could be "inoculated" against new attacks.[9]

This kind of language is not merely picturesque anthropomorphism. It reflects a profound shift in perception about the relationship of computers to humans. If computers can become ill, then they can become healthy. Once healthy, they can think clearly and make decisions. The computer, it is implied, has a will, has intentions, has reasons—which means that humans are relieved of responsibility for the computer's decisions. Through a curious form of grammatical alchemy, the sentence "We use the computer to calculate" comes to mean "The computer calculates." If a computer calculates, then it may decide to miscalculate or not calculate at all. That is what bank tellers mean when they tell you that they cannot say how much money is in your checking account because "the computers are down." The implication, of course, is that no person at the bank is responsible. Computers make mistakes or get tired or become ill. Why blame people? We may call this line of thinking an "agentic shift," a term I borrow from Stanley Milgram to name the process whereby humans transfer responsibility for an outcome from themselves to a more abstract agent.[10] When this happens, we have relinquished control, which in the case of the computer means that we may, without excessive remorse, pursue ill-advised or even inhuman goals because the computer can accomplish them or be imagined to accomplish them.

Machines of various kinds will sometimes assume a human or, more likely, a superhuman aspect. Perhaps the most absurd case I know of is in a remark a student of mine once made on a sultry summer day in a room without air conditioning. On being told the thermometer read ninety-eight degrees Fahrenheit, he replied, "No wonder it's so hot!" Nature was off the hook. If only the thermometers would behave themselves, we could be comfortable. But computers are far more "human" than thermometers or almost any other kind of technology. Unlike most machines, computers do no work; they direct work. They are, as Norbert Wiener said, the technology of "command and control" and have little value without something to control. This is why they are of such importance to bureaucracies.

Naturally, bureaucrats can be expected to embrace a technology that helps to create the illusion that decisions are not under their control. Because of its seeming intelligence and impartiality, a computer has an almost magical tendency to direct attention away from the people in charge of bureaucratic functions and toward itself, as if the computer were the true source of authority. A bureaucrat armed with a computer is the unacknowledged legislator of our age, and a terrible burden to bear. We cannot dismiss the possibility that, if Adolf Eichmann had been able to say that it was not he but a battery of computers that directed the Jews to the appropriate crematoria, he might never have been asked to answer for his actions.

Although (or perhaps because) I came to "administration" late in my academic career, I am constantly amazed at how obediently people accept explanations that begin with the words "The computer shows . . ." or "The computer has determined . . ." It is Technopoly's equivalent of the sentence "It is God's will," and the effect is roughly the same. You will not be surprised to know that I rarely resort to such humbug. But on occasion, when pressed to the wall, I have yielded. No one has as yet replied, "Garbage in, garbage out." Their defenselessness has something Kafkaesque about it. In *The Trial,* Josef K. is charged with a crime—of what nature, and by whom the charge is made, he does not know. The computer turns too many of us into Josef Ks. It often functions as a kind of impersonal accuser which does not reveal, and is not required to reveal, the sources of the judgments made against us. It is apparently sufficient that the computer has pronounced. Who has put the data in, for what purpose, for whose convenience, based on what assumptions are questions left unasked.

This is the case not only in personal matters but in public decisions as well. Large institutions such as the Pentagon, the Internal Revenue Service, and multinational corporations tell us that their decisions are made on the basis of solutions generated by computers, and this is usually good enough to put our minds at ease or, rather, to sleep. In any case, it constrains us from making complaints or accusations. In part for this reason, the computer has strengthened bureaucratic institutions and suppressed the impulse toward significant social change. "The arrival of the Computer Revolution and the founding of the Computer Age have been announced many times," Weizenbaum has written. "But if the triumph of a revolution is to be measured in terms of the social revision it entrained, then there has been no computer revolution."[11]

In automating the operation of political, social, and commercial enterprises, computers may or may not have made them more efficient but they have certainly diverted attention from the question whether or not such enterprises are necessary or how they might be improved. A university, a political party, a religious denomination, a judicial proceeding, even corporate board meetings are not improved by automating their operations. They are made more imposing, more technical, perhaps more authoritative, but defects in their assumptions, ideas, and theories will remain untouched. Computer technology, in other words, has not yet come close to the printing press in its power to generate radical and substantive social, political, and religious thought. If the press was, as David Riesman called it, "the gunpowder of the mind," the computer, in its capacity to smooth over unsatisfactory institutions and ideas, is the talcum powder of the mind.

I do not wish to go as far as Weizenbaum in saying that computers are merely ingenious devices to fulfill unimportant functions and that the computer revolution is an explosion of nonsense. Perhaps that judgment will be in need of amendment in the future, for the computer is a technology of a thousand uses—the Proteus of machines, to use Seymour Papert's phrase. One must note, for example, the use of computer-generated images in the phenomenon known as Virtual Reality. Putting on a set of miniature goggle-mounted screens, one may block out the real world and move through a simulated three-dimensional world which changes its components with every movement of one's head. That Timothy Leary is an enthusiastic proponent of Virtual Reality does not suggest that there is a constructive future for this device. But who knows? Perhaps, for those who can no longer cope with the real world, Virtual Reality will provide better therapy than ELIZA.

What is clear is that, to date, computer technology has served to strengthen Technopoly's hold, to make people believe that technological innovation is synonymous with human progress. And it has done so by advancing several interconnected ideas.

It has, as already noted, amplified beyond all reason the metaphor of machines as humans and humans as machines. I do not claim, by the way, that computer technology originated this metaphor. One can detect it in medicine, too: doctors and patients have come to believe that, like a machine, a human being is made up of parts which when defective can be replaced by mechanical parts that function as the original did without impairing or even affecting any other part of the machine. Of course, to some degree that assumption works, but since a human being is in fact not a machine but a biological organism all of whose organs are interrelated and profoundly affected by mental states, the human-as-machine metaphor has serious medical limitations and can have devastating effects. Something similar may be said of the mechanistic metaphor when applied to workers. Modern industrial techniques are made possible by the idea that a machine is made up of isolatable and interchangeable parts. But in organizing factories so that workers are also conceived of as isolatable and interchangeable parts, industry has engendered deep alienation and bitterness. This was the point of Charlie Chaplin's *Modern Times,* in which he tried to show the psychic damage of the metaphor carried too far. But because the computer "thinks" rather than works, its power to energize mechanistic metaphors is unparalleled and of enormous value to Technopoly, which depends on our believing that we are at our best when acting like machines, and that in significant ways machines may be trusted to act as our surrogates. Among the implications of these beliefs is a loss of confidence in human judgment and subjectivity. We have devalued the singular human capacity to see things whole in all their psychic, emotional and moral dimensions, and we have replaced this with faith in the powers of technical calculation.

Because of what computers commonly do, they place an inordinate emphasis on the technical processes of communication and offer very little in the way of substance. With the exception of the electric light, there never has been a technology that better exemplifies Marshall McLuhan's aphorism "The medium is the message." The computer is almost all process. There are, for example, no "great computerers," as there are great writers, painters, or musicians. There are "great programs" and "great programmers," but their greatness lies in their ingenuity either in simulating a human function or in creating new possibilities of calculation, speed, and volume.[12] Of course, if J. David Bolter is right, it is possible that in the future computers will emerge as a new kind of book, expanding and enriching the tradition of writing technologies.[13] Since printing created new forms of literature when it replaced the handwritten manuscript, it is possible that electronic writing will do the same. But for the moment, computer technology functions more as a new mode of transportation than as a new means of substantive communication. It moves information—lots of it, fast, and mostly in a calculating mode. The computer, in fact, makes possible the fulfillment of Descartes' dream of the mathematization of the world. Computers make it easy to convert facts into statistics and to translate problems into equations. And whereas this can be useful (as when the process reveals a pattern that would otherwise go unnoticed), it is diversionary and dangerous when applied indiscriminately to human affairs. So is the computer's emphasis on speed and especially its capacity to generate and store unprecedented quantities of information. In specialized contexts, the value of calculation, speed, and voluminous information may go uncontested. But the "message" of computer technology is comprehensive and domineering. The computer argues, to put it baldly, that the most serious problems confronting us at both personal and public levels require technical solutions through fast access to information otherwise unavailable. I would argue that this is, on the face of it, nonsense. Our most serious problems are not technical, nor do they arise from inadequate information. If a nuclear catastrophe occurs, it shall not be because of inadequate information. Where people are dying of starvation, it does not occur because of inadequate information. If families break up, children are mistreated, crime terrorizes a city, education is impotent, it does not happen because of inadequate information. Mathematical equations, instantaneous communication, and vast quantities of information have nothing whatever to do with any of these problems. And the computer is useless in addressing them.

And yet, because of its "universality," the computer compels respect, even devotion, and argues for a comprehensive role in all fields of human activity. Those who insist that it is foolish to deny the computer vast sovereignty are singularly devoid of what Paul Goodman once called "technological modesty"—that is, having a sense of the whole and not claiming or obtruding more than a particular function warrants. Norbert Wiener warned about lack of modesty when he remarked that, if digital computers had been in common use before the atomic bomb was invented, people would have said that the bomb could not have been invented without computers. But it was. And it is important to remind ourselves of how many things are quite possible to do without the use of computers.

Seymour Papert, for example, wishes students to be epistemologists, to think critically, and to learn how to create knowledge. In his book *Mindstorms,* he gives the impression that his computer program known as LOGO now makes this possible. But good teachers have been doing this for centuries without the benefit of LOGO. I do not say that LOGO, when used properly by a skilled teacher, will not help, but I doubt that it can do better than pencil and paper, or speech itself, when used properly by a skilled teacher.

When the Dallas Cowboys were consistently winning football championships, their success was attributed to the fact that computers were used to evaluate and select team members. During the past several years, when Dallas has been hard put to win more than a few games, not much has been said about the computers, perhaps because people have realized that computers have nothing to do with winning football games, and never did. One might say the same about writing lucid, economical, stylish prose, which has nothing to do with word-processors. Although my students don't believe it, it is actually possible to write well without a processor and, I should say, to write poorly with one.

Technological immodesty is always an acute danger in Technopoly, which encourages it. Technopoly also encourages insensitivity to what skills may be lost in the acquisition of new ones. It is important to remember what can be done without computers, and it is also important to remind ourselves of what may be lost when we do use them.

I have before me an essay by Sir Bernard Lovell, founder of Britain's Jodrell Bank Observatory, in which he claims that computers have stifled scientific creativity.[14] After writing of his awe at the ease with which computerized operations provide amazing details of distant galaxies, Sir Bernard expresses concern that "literal-minded, narrowly focused computerized research is proving antithetical to the free exercise of that happy faculty known as serendipity—that is, the knack of achieving favorable results more or less by chance." He proceeds to give several examples of monumental but serendipitous discoveries, contends that there has been a dramatic cessation of such discoveries, and worries that computers are too narrow as filters of information and therefore may be antiserendipitous. He is, of course, not "against" computers, but is merely raising questions about their costs.

Dr. Clay Forishee, the chief FAA scientist for human performance issues, did the same when he wondered whether the automated operation of commercial aircraft has not disabled pilots from creatively responding when something goes wrong. Robert Buley, flight-standards manager of Northwest Airlines, goes further. He is quoted as saying, "If we have human operators subordinated to technology then we're going to lose creativity [in emergencies]." He is not "against" computers. He is worried about what we lose by using them.[15]

M. Ethan Katsch, in his book *The Electronic Media and the Transformation of Law,* worries as well. He writes, "The replacement of print by computerized systems is promoted to the legal profession simply as a means to increase efficiency."[16] But he goes on to say that, in fact, the almost unlimited capacity of computers to store and retrieve information threatens the authority of precedent, and he adds that the threat is completely unrecognized. As he notes, "a system of precedent is unnecessary when there are very few accessible cases, and unworkable when there are too many." If this is true, or even partly true, what exactly does it mean? Will lawyers become incapable of choosing relevant precedents? Will judges be in constant confusion from "precedent overload"?

We know that doctors who rely entirely on machinery have

lost skill in making diagnoses based on observation. We may well wonder what other human skills and traditions are being lost by our immersion in a computer culture. Technopolists do not worry about such things. Those who do are called technological pessimists, Jeremiahs, and worse. I rather think they are imbued with technological modesty, like King Thamus.

8

Invisible Technologies

If we define ideology as a set of assumptions of which we are barely conscious but which nonetheless directs our efforts to give shape and coherence to the world, then our most powerful ideological instrument is the technology of language itself. Language is pure ideology. It instructs us not only in the names of things but, more important, in what things can be named. It divides the world into subjects and objects. It denotes what events shall be regarded as processes, and what events, things. It instructs us about time, space, and number, and forms our ideas of how we stand in relation to nature and to each other. In English grammar, for example, there are always subjects who act, and verbs which are their actions, and objects which are acted upon. It is a rather aggressive grammar, which makes it difficult for those of us who must use it to think of the world as benign. We are obliged to know the world as made up of things pushing against, and often attacking, one another.

Of course, most of us, most of the time, are unaware of how language does its work. We live deep within the boundaries of our linguistic assumptions and have little sense of how the world looks to those who speak a vastly different tongue. We tend to assume that everyone sees the world in the same way, irrespective of differences in language. Only occasionally is this illusion challenged, as when the differences between linguistic ideologies become noticeable by one who has command over two languages that differ greatly in their structure and history. For example, several years ago, Susumu Tonegawa, winner of the 1987 Nobel Prize in Medicine, was quoted in the newspaper *Yomiuri* as saying that the Japanese language does not foster clarity or effective understanding in scientific research. Addressing his countrymen from his post as a professor at MIT in Cambridge, Massachusetts, he said, "We should consider changing our thinking process in the field of science by trying to reason in English." It should be noted that he was not saying that English is better than Japanese; only that English is better than Japanese for the purposes of scientific research, which is a way of saying that English (and other Western languages) have a particular ideological bias that Japanese does not. We call that ideological bias "the scientific outlook." If the scientific outlook seems natural to you, as it does to me, it is because our language makes it appear so. What we think of as reasoning is determined by the character of our language. To reason in Japanese is apparently not the same thing as to reason in English or Italian or German.

To put it simply, like any important piece of machinery—television or the computer, for example—language has an ideological agenda that is apt to be hidden from view. In the case of language, that agenda is so deeply integrated into our personalities and world-view that a special effort and, often, special training are required to detect its presence. Unlike television or the computer, language appears to be not an extension of our powers but simply a natural expression of who and what we are. This is the great secret of language: Because it comes from inside us, we believe it to be a direct, unedited, unbiased, apolitical expression of how the world really is. A machine, on the other hand, is outside of us, clearly created by us, modifiable by us, even discardable by us; it is easier to see how a machine re-creates the world in its own image. But in many respects, a sentence functions very much like a machine, and this is nowhere more obvious than in the sentences we call questions.

As an example of what I mean, let us take a "fill-in" question, which I shall require you to answer exactly if you wish full credit:

Thomas Jefferson died in the year ——.

Suppose we now rephrase the question in multiple-choice form:

Thomas Jefferson died in the year (a) 1788 (b) 1826 (c) 1926 (d) 1809.

Which of these two questions is easier to answer? I assume you will agree with me that the second question is easier unless you happen to know precisely the year of Jefferson's death, in which case neither question is difficult. However, for most of us who know only roughly when Jefferson lived, Question Two has arranged matters so that our chances of "knowing" the answer are greatly increased. Students will always be "smarter" when answering a multiple-choice test than when answering a "fill-in" test, even when the subject matter is the same. A question, even of the simplest kind, is not and can never be unbiased. I am not, in this context, referring to the common accusation that a particular test is "culturally biased." Of course questions can be culturally biased. (Why, for example, should anyone be asked about Thomas Jefferson at all, let alone when he died?) My purpose is to say that the structure of any question is as devoid of neutrality as is its content. The form of a question may ease our way or pose obstacles. Or, when even slightly altered, it may generate antithetical answers, as in the case of the two priests who, being unsure if it was permissible to smoke and pray at the same time, wrote to the Pope for a definitive answer. One priest phrased the question "Is it permissible to smoke while praying?" and was told it is not, since prayer should be the focus of one's whole attention; the other priest asked if it is permissible to pray while smoking and was told that it is, since it is always appropriate to pray. The form of a question may even block us from seeing solutions to problems that become visible through a different question. Consider the following story, whose authenticity is questionable but not, I think, its point:

Once upon a time, in a village in what is now Lithuania, there arose an unusual problem. A curious disease afflicted many of the townspeople. It was mostly fatal (though not always), and its onset was signaled by the victim's lapsing into a deathlike coma. Medical science not being quite so advanced as it is now, there was no definite way of knowing if the victim was actually dead when burial appeared seemly. As a result, the townspeople feared that several of their relatives had already been buried alive and that a similar fate might await them. How to overcome this uncertainty was their dilemma.

One group of people suggested that the coffins be well stocked with water and food and that a small air vent be drilled into them, just in case one of the "dead" happened to be alive. This was expensive to do but seemed more than worth the trouble. A second group, however, came up with a less expensive and more efficient idea. Each coffin would have a twelve-inch stake affixed to the inside of the coffin lid, exactly at the level of the heart. Then, when the coffin was closed, all uncertainty would cease.

The story does not indicate which solution was chosen, but for my purposes the choice is irrelevant. What is important to note is that different solutions were generated by different questions. The first solution was an answer to the question, How can we make sure that we do not bury people who are still alive? The second was an answer to the question, How can we make sure that everyone we bury is dead?

Questions, then, are like computers or television or stethoscopes or lie detectors, in that they are mechanisms that give direction to our thoughts, generate new ideas, venerate old ones, expose facts, or hide them. In this chapter, I wish to consider mechanisms that act like machines but are not normally thought of as part of Technopoly's repertoire. I must call attention to them precisely because they are so often overlooked. For all practical purposes, they may be considered technologies—technologies in disguise, perhaps, but technologies all the same.

Aside from language itself, I don't suppose there is a clearer example of a technology that doesn't look like one than the mathematical sign known as zero. A brief word about it may help to illuminate later examples.

The zero made its way from India to Europe in the tenth century. By the thirteenth century, it had taken hold of Western consciousness. (It was unknown to the Romans and the classical Greeks, although analogous concepts were known to Babylonian mathematicians of the Hellenistic period.) Without the zero, you will find it difficult to perform any of the calculations that are quite simple to do with it. If you should try multiplying MMMMMM by MMDCXXVI, you will have this point confirmed. I have been told, by the way, that such a calculation *can* be done, but the process is so laborious that the task is unlikely to be completed, a truth that did not escape the notice of medieval mathematicians. There is, in fact, no evidence that Roman numerals were ever used, or intended to be used, for calculation. For that purpose, mathematicians used an abacus, and between the tenth and thirteenth centuries, a struggle of sorts took place between abacists, who wrote Roman numerals but calculated with the abacus, and algorists, who used Hindu numerals employing the zero sign. The objection raised by the abacists was that the zero registered the *absence* of a power of ten, which no Roman numeral did, and which struck them as philosophically and perhaps aesthetically offensive. After all, the zero is a sign that affects values of numerals wherever it occurs but has no value in itself. It is a sign about signs, whose very etymology, via "cipher" from the Hindu word for "void," suggests the idea of "nothingness." To the abacists, it was a bizarre idea to have a sign marking "nothing," and I fear that I would have sided with the abacists.

I speak of the zero for two reasons: First, to underscore that it is a kind of technology that makes both possible and easy certain kinds of thoughts which, without it, would remain inaccessible to the average person. If it does not exactly have an ideology, it contains, at least, an idea. I have previously alluded to the technology of using letters or numbers to grade students' papers, and to the Greek discovery of the technology of alphabetization: like the use of zero, these are examples of how symbols may function like machines in creating new mind-sets and therefore new conceptions of reality. Second, the use of the zero and, of course, the Hindu numbering system of which it was a part made possible a sophisticated mathematics which, in turn, led to one of the most powerful technologies now in use: statistics.

Statistics makes possible new perceptions and realities by making visible large-scale patterns. Its uses in science are too well known to warrant notice here, except to remark that if, as the physicists tell us, the world is made up of probabilities at the level of subatomic particles, then statistics is the only means by which to describe its operations. Indeed, the uncertainty principle ensures that in the nature of things physics is unable to do more than make statistical predictions.

Of course, it is possible that physicists conceive of the world as probabilistic *because* statistics was invented. But that is not the question I wish to pursue here. A more practical question is, To what extent has statistics been allowed entry to places where it does not belong? Technopoly, by definition, grants free rein to any technology, and we would expect that no limits have been placed on the use of statistics. We would expect correctly.

Perhaps the most abusive example is found in the work of Francis Galton, who was born in 1822, died in 1911, and therefore lived during the richest period of technological invention. He may be thought of as one of the Founding Fathers of Technopoly. Galton is also known as the founder of "eugenics," a term he coined, which means the "science" of arranging marriage and family so as to produce the best possible offspring based on the hereditary characteristics of the parents. He believed that anything could be measured and that statistical procedures, in particular, were the technology that could open the pathway to real knowledge about every form of human behavior. The next time you watch a televised beauty contest in which women are ranked numerically, you should remember Francis Galton, whose pathological romance with numbers originated this form of idiocy. Being unsatisfied with vagueness about where the most "beauty" was to be found, he constructed a "beauty map" of the British Isles. As he told us, he classified "the girls I passed in streets or elsewhere as attractive, indifferent, or repellent." He then proved statistically that London had the most beautiful girls, Aberdeen the ugliest; this no doubt made it awkward for Galton to spend his vacation in Scotland. If this were not enough, he also invented a method for quantifying boredom (by counting the number of fidgets) and even proposed a statistical inquiry for determining the efficacy of prayer.

But Galton's main interest was in demonstrating, statistically, the inheritance of intelligence. To that end, he established a laboratory at the International Exposition of 1884, where for threepence people could have their skulls measured and receive Galton's assessment of their intelligence. Apparently, a visitor received no extra credit for demanding his or her money back, which would surely have been a sign of intelligence. We can be sure that not many did, since Galton was considered a major intellect of his day. In fact, Lewis Terman, the man most responsible for promoting IQ tests in America, calculated that Galton's IQ was more than 200. Terman, who fancied making such estimates of the dead, ranked Charles Darwin (Galton's cousin, incidentally) at a mere 135, and poor Copernicus somewhere between 100 and 110.[1]

For a definitive history and analysis of the malignant role played by statistics in the "measurement" of intelligence, I refer the reader to Stephen Jay Gould's brilliant book *The Mismeasure of Man.* Here, I will only cite three points made by Gould, which I believe are sufficient to convince anyone with a higher IQ than Copernicus of the dangers of abusing statistics.

The first problem is called reification, which means converting an abstract idea (mostly, a word) into a thing. In this context, reification works in the following way: We use the word "intelligence" to refer to a variety of human capabilities of which we approve. There is no such thing as "intelligence." It is a word, not a thing, and a word of a very high order of abstraction. But if we believe it to be a thing like the pancreas or liver, then we will believe scientific procedures can locate it and measure it.

The second problem is ranking. Ranking requires a criterion for assigning individuals to their place in a single series. As Gould remarks, what better criterion can be used than an objective number? In the ranking of intelligence, we therefore assume that intelligence is not only a thing, but a single thing, located in the brain, and accessible to the assignment of a number. It is as if "beauty" were determined to inhere in the size of a woman's breasts. Then all we would have to do is measure breasts and rank each woman accordingly, and we would have an "objective" measure of "beauty."

The third point is that in doing this, we would have formulated our question "Who is the fairest of all?" in a restricted and biased way. And yet this would go unnoticed, because, as Gould writes, "The mystique of science proclaims that numbers are the ultimate test of objectivity." This means that the way we have defined the concept will recede from our consciousness—that is, its fundamental subjectivity will become invisible, and the objective number itself will become reified. One would think that such a process would appear ridiculous on the breast of it, especially since, by believing it, we must conclude that Dolly Parton is objectively proved to be more beautiful than Audrey Hepburn. Or, in the case of intelligence, that Galton had twice as much of it as Copernicus.

Nonetheless, in Technopoly all this is taken very seriously, albeit not without a few protests. After a lifetime of working in the field of intelligence measurement, E. L. Thorndike observed that intelligence tests suffer from three small defects: "Just what they measure is not known; how far it is proper to add, subtract, multiply, divide, and compute ratios with the measures obtained is not known; just what the measures signify concerning intellect is not known."[2] In other words, those who administer intelligence tests quite literally do not know what they are doing. That is why David McClelland remarked, "Psychologists should be ashamed of themselves for promoting a view of general intelligence that has engendered such a testing program." Joseph Weizenbaum summed it up by saying, "Few 'scientific' concepts have so thoroughly muddled the thinking of both scientists and the general public as that of the 'intelligence quotient' or 'IQ.' The idea that intelligence can be quantitatively measured along a single linear scale has caused untold harm to our society in general, and to education in particular."[3]

Gould has documented some of this harm, and Howard Gardner has tried to alleviate it (in his book *Frames of Mind*). But Technopoly resists such reproaches, because it needs to believe that science is an entirely objective enterprise. Lacking a lucid set of ethics and having rejected tradition, Technopoly searches for a source of authority and finds it in the idea of statistical objectivity.

This quest is especially evident not only in our efforts to determine precisely how smart people are but also in our attempts to find out precisely how smart *groups* of people are. Aside from the fact that the procedures used do not and *cannot* give such an answer, one must ask, Of what earthly use is it to declare that one group of people is smarter than another? Suppose it is shown that according to objective measures Asians have more "intelligence" than Caucasians, or that Caucasians have more than African-Americans. Then what? Of what use is this information to, say, a teacher or an employer? Is the teacher or employer to assume that a particular Asian is smarter than a particular African-American? Or even that six Asians are smarter than six African-Americans? Obviously not. And yet who knows? We must keep in mind the story of the statistician who drowned while trying to wade across a river with an average depth of four feet. That is to say, in a culture that reveres statistics, we can never be sure what sort of nonsense will lodge in people's heads.

The only plausible answer to the question why we use statistics for such measurements is that it is done for sociopolitical reasons whose essential malignancy is disguised by the cover of "scientific inquiry." If we believe that blacks are dumber than whites, and that this is not merely our opinion but is confirmed by objective measures, then we can believe we have an irreproachable authority for making decisions about the allocation of resources. This is how, in Technopoly, science is used to make democracy "rational."

Polling is still another way. Just as statistics has spawned a huge testing industry, it has done the same for the polling of "public opinion." One may concede, at the start, that there are some uses of polling that may be said to be reliable, especially if the case involves a greatly restricted question such as, Do you plan to vote for X or Y? But to say a procedure is reliable is not to say it is useful. The question is as yet undecided whether knowledge of voter trends during a political campaign enriches or demeans the electoral process. But when polls are used to guide public policy, we have a different sort of issue altogether.

I have been in the presence of a group of United States congressmen who were gathered to discuss, over a period of two days, what might be done to make the future of America more survivable and, if possible, more humane. Ten consultants were called upon to offer perspectives and advice. Eight of them were pollsters. They spoke of the "trends" their polling uncovered; for example, that people were no longer interested in the women's movement, did not regard environmental issues as of paramount importance, did not think the "drug problem" was getting worse, and so on. It was apparent, at once, that these polling results would become the basis of how the congressmen thought the future should be managed. The ideas the congressmen had (all men, by the way) receded to the background. Their own perceptions, instincts, insights, and experience paled into irrelevance. Confronted by "social scientists," they were inclined to do what the "trends" suggested would satisfy the populace.[4]

It is not unreasonable to argue that the polling of public opinion puts democracy on a sound and scientific footing. If our political leaders are supposed to represent us, they must have some information about what we "believe." In principle, there is no problem here. The problems lie elsewhere, and there are at least four of them.

The first has to do with the forms of the questions that are put to the public. I refer the reader to the matter of whether it is proper to smoke and pray at the same time. Or, to take a more realistic example: If we ask people whether they think it acceptable for the environment to continue to be polluted, we are likely to come up with answers quite different from those generated by the question, Do you think the protection of the environment is of paramount importance? Or, Do you think safety in the streets is more important than environmental protection? The public's "opinion" on almost any issue will be a function of the question asked. (I might point out that in the seminar held by the congressmen, not one asked a question about the questions. They were interested in results, not in how these were obtained, and it did not seem to occur to them that the results and how they are obtained are inseparable.)

Typically, pollsters ask questions that will elicit yes or no answers. Is it necessary to point out that such answers do not give a robust meaning to the phrase "public opinion"? Were you, for example, to answer "No" to the question "Do you think the drug problem can be reduced by government programs?" one would hardly know much of interest or value about your opinion. But allowing you to speak or write at length on the matter would, of course, rule out using statistics. The point is that the use of statistics in polling changes the meaning of "public opinion" as dramatically as television changes the meaning of "political debate." In the American Technopoly, public opinion is a yes or no answer to an unexamined question.

Second, the technique of polling promotes the assumption that an opinion is a thing inside people that can be exactly located and extracted by the pollster's questions. But there is an alternative point of view, of which we might say, it is what Jefferson had in mind. An opinion is not a momentary thing but a process of thinking, shaped by the continuous acquisition of knowledge and the activity of questioning, discussion, and debate. A question may "invite" an opinion, but it also may modify and recast it; we might better say that people do not exactly "have" opinions but are, rather, involved in "opinioning." That an opinion is conceived of as a measurable thing falsifies the process by which people, in fact, do their opinioning; and how people do their opinioning goes to the heart of the meaning of a democratic society. Polling tells us nothing about this, and tends to hide the process from our view.

Which leads to the third point. Generally, polling ignores what people know about the subjects they are queried on. In a culture that is not obsessed with measuring and ranking things, this omission would probably be regarded as bizarre. But let us imagine what we would think of opinion polls if the questions came in pairs, indicating what people "believe" and what they "know" about the subject. If I may make up some figures, let us suppose we read the following: "The latest poll indicates that 72 percent of the American public believes we should withdraw economic aid from Nicaragua. Of those who expressed this opinion, 28 percent thought Nicaragua was in central Asia, 18 percent thought it was an island near New Zealand, and 27.4 percent believed that 'Africans should help themselves,' obviously confusing Nicaragua with Nigeria. Moreover, of those polled, 61.8 percent did not know that we give economic aid to Nicaragua, and 23 percent did not know what 'economic aid' means." Were pollsters inclined to provide such information, the prestige and power of polling would be considerably reduced. Perhaps even congressmen, confronted by massive ignorance, would invest their own understandings with greater trust.

The fourth problem with polling is that it shifts the locus of responsibility between political leaders and their constituents. It is true enough that congressmen are supposed to represent the interests of their constituents. But it is also true that congressmen are expected to use their own judgment about what is in the public's best interests. For this, they must consult their own experience and knowledge. Before the ascendance of polling, political leaders, though never indifferent to the opinions of their constituents, were largely judged on their capacity to make decisions based on such wisdom as they possessed; that is, political leaders were responsible for the decisions they made. With the refinement and extension of the polling process, they are under increasing pressure to forgo deciding anything for themselves and to defer to the opinions of the voters, no matter how ill-informed and shortsighted those opinions might be.

We can see this process of responsibility-shift even more clearly in the case of the statistically based ratings of television shows. The definition of a "good" television show has become purely and simply a matter of its having high ratings. A "bad" show has low ratings. The responsibility of a television writer, therefore, begins and ends with his or her ability to create a show that many millions of viewers will watch. The writer, in a word, is entirely responsible to the audience. There is no need for the writer to consult tradition, aesthetic standards, thematic plausibility, refinements of taste, or even plain comprehensibility. The iron rule of public opinion is all that matters. Television executives are fond of claiming that their medium is the most democratic institution in America: a plebiscite is held every week to determine which programs will survive. This claim is given added weight by a second claim: creative artists have never been indifferent to the preferences and opinions of their audiences. Writers, for example, write *for* people, for their approbation and understanding. But writers also write for themselves and because they have something they want to say, not always because readers have something they want to hear. By giving constant deference to public preferences, polling changes the motivation of writers; their entire effort is to increase "the numbers." Popular literature now depends more than ever on the wishes of the audience, not the creativity of the artist.

Before leaving the subject of the technology of statistics, I must call attention to the fact that statistics creates an enormous amount of completely useless information, which compounds the always difficult task of locating that which is useful to a culture. This is more than a case of "information-overload." It is a matter of "information-trivia," which has the effect of placing all information on an equal level. No one has expressed this misuse of a technology better than the *New Yorker* magazine cartoonist Mankoff. Showing an attentive man watching television news, Mankoff has the newscaster saying, "A preliminary census report indicates that for the first time in our nation's history female anthropologists outnumber male professional golfers." When statistics and computers are joined, volumes of garbage are generated in public discourse. Those who have watched television sports programs will know that Mankoff's cartoon is, in fact, less of a parody than a documentary. Useless, meaningless statistics flood the attention of the viewer. Sportscasters call them "graphics" in an effort to suggest that the information, graphically presented, is a vital supplement to the action of the game. For example: "Since 1984, the Buffalo Bills have won only two games in which they were four points ahead with less than six minutes to play." Or this: "In only 17 percent of the times he has pitched at Shea Stadium has Dwight Gooden struck out the third and fourth hitters less than three times when they came to bat with more than one runner on base."[5] What is one to do with this or to make of it? And yet there seems to be a market for useless information. Those who read *USA Today*, for example, are offered on the front page of each issue an idiotic statistic of the day that looks something like this: "The four leading states in banana consumption from 1980 through 1989 are Kansas, North Dakota, Wyoming, and Louisiana. Oddly, Nevada, which was ninth in 1989, fell to twenty-sixth last year, which is exactly where it ranks in kiwi consumption."[6]

It is surprising how frequently such blather will serve as the backbone of conversations which are essentially meaningless. I have heard New Yorkers, with a triumphant flourish, offer out-of-towners the statistic that New York City is only eighth in the nation in per-capita violent crimes and then decline to go outside because it was past 6:00 p.m.

I do not say, of course, that all such statistical statements are useless. If we learn that one out of every four black males between the ages of twenty and thirty has spent some time in prison, and that the nation's expenditure for the education of black children is 23 percent less than it is for white children, we may have some statistical facts that will help us to see a cause-and-effect relationship, and thereby suggest a course of action. But statistics, like any other technology, has a tendency to run out of control, to occupy more of our mental space than it warrants, to invade realms of discourse where it can only wreak havoc. When it is out of control, statistics buries in a heap of trivia what is necessary to know.

And there is another point, which in fact is the core of this chapter. Some technologies come in disguise. Rudyard Kipling called them "technologies in repose." They do not look like technologies, and because of that they do their work, for good or ill, without much criticism or even awareness. This applies not only to IQ tests and to polls and to all systems of ranking and grading but to credit cards, accounting procedures, and achievement tests. It applies in the educational world to what are called "academic courses," as well. A course is a technology for learning. I have "taught" about two hundred of them and do not know why each one lasts exactly fifteen weeks, or why each meeting lasts exactly one hour and fifty minutes. If the answer is that this is done for administrative convenience, then a course is a fraudulent technology. It is put forward as a desirable structure for learning when in fact it is only a structure for allocating space, for convenient record-keeping, and for control of faculty time. The point is that the origin of and raison d'être for a course are concealed from us. We come to believe it exists for one reason when it exists for quite another. One characteristic of those who live in a Technopoly is that they are largely unaware of both the origins and the effects of their technologies.[7]

Perhaps the most interesting example of such lack of awareness is the widespread belief that modern business invented the technology of management. Management is a system of power and control designed to make maximum use of relevant knowledge, the hierarchical organization of human abilities, and the flow of information from bottom to top and back again. It is generally assumed that management was created by business enterprises as a rational response to the economic and technological demands of the Industrial Revolution. But research by Alfred Chandler, Sidney Pollard, and especially Keith Hoskin and Richard Macve reveals a quite different picture and leads to a startling conclusion: modern business did not invent management; management invented modern business.[8]

The most likely place for management to have originated is, of course, in Great Britain in the late eighteenth and early nineteenth centuries. But there is no evidence that British industry knew anything about management as late as 1830, nor did there exist anything approximating a "managerial class." Management was created in the United States "out of the blue," as Hoskin and Macve say. It was not a creation of any obvious needs of American industry, which was only a marginal force in the world economy in the mid-nineteenth century. The roots of management may be traced to a new educational system, introduced in 1817 to the United States Military Academy by the academy's fourth superintendent, Sylvanus Thayer. Thayer made two innovations. The first, borrowed from the Ecole Polytechnique in Paris, was to grade examinations by giving numerical marks. As I have previously noted, the grading of student papers originated in Cambridge University toward the end of the eighteenth century, and the practice was taken up by several schools on the Continent. Thayer's use of this technology is probably the first instance of it in America. As every teacher knows, the numerical mark changes the entire experience and meaning of learning. It introduces a fierce competition among students by providing sharply differentiated symbols of success and failure. Grading provides an "objective" measure of human performance and creates the unshakable illusion that accurate calculations can be made of worthiness. The human being becomes, to use Michel Foucault's phrase, "a calculable person."

Thayer's second innovation, apparently his own invention, was a line-and-staff system. He divided the academy into two divisions, each organized hierarchically. As Hoskin and Macve describe it: "Daily, weekly and monthly reports were required, all in writing. There were continual relays of written communication and command, going from the bottom to the top of each line, before being consolidated and passed to the central 'Staff Office.' " Thayer rejected the traditional leader's role of direct, visible command. He ruled indirectly through the medium of written reports, charts, memos, personnel files, etc., not unlike the way a modern CEO functions.

We do not know how most of the two hundred cadets at the academy reacted to Thayer's new system (which Hoskin and Macve term the "grammatocentric principle," meaning that everything was organized around the use of writing). But we do know that two of them, Daniel Tyler and George Whistler, were impressed. Both were in the graduating class of 1819, and took with them their lieutenant's rank and Thayer's general approach to organizations.

Daniel Tyler, working at the Springfield Armory, did a time-and-motion study in 1832 (sixty years before Frederick Taylor's "scientific management" got under way) and established objectively based norms of production for every job in the armory. Workers were kept under surveillance, and their actual productivity was measured against the established productivity norms. Tyler also introduced quality control and inventory accounting. The result of all these methods was a dramatic increase in productivity and decrease in costs.

Meanwhile, George Whistler (incidentally, the father of James Whistler and therefore the husband of "Whistler's Mother"), having become the chief engineer of the Western Railroad, developed a managerial system in 1839 that would have made Sylvanus Thayer proud. He organized the railroad along hierarchical lines, beginning with a central staff office, descending to regional managers and then local managers. He employed, to great effect, the grammatocentric principle, which he had no doubt learned well at the academy when serving in the staff office as cadet staff sergeant major.

The principles of calculability and grammatocentrism are, of course, the foundation of modern systems of management. Calculability led inevitably to such ideas as detailed accounting systems, inventory control, and productivity norms. Grammatocentrism promoted the idea that the best way to run a business is to know it through reports of those lower down the line. One manages, in other words, by the "numbers" and by being removed from the everyday realities of production.

It is worth saying that the basic structure of business management originated in nonbusiness contexts. Still, it did not take very long for American businesses to begin to adopt the principles of Thayer, Tyler, and Whistler, and by doing so they created what we now think of as a modern corporation. Indeed, management defines what we mean by a corporation, and has led John Kenneth Galbraith to remark in *The New Industrial State:* "More perhaps than machinery, massive and complex business organizations are the tangible manifestation of advanced technology."

There are two reasons why the case of management is instructive. First, as suggested by Galbraith, management, like the zero, statistics, IQ measurement, grading papers, or polling, functions as does any technology. It is not made up of mechanical parts, of course. It is made up of procedures and rules designed to standardize behavior. We may call any such system of procedures and rules a technique; and there is nothing to fear from techniques, unless, like so much of our machinery, they become autonomous. There's the rub. In a Technopoly, we tend to believe that only through the autonomy of techniques (and machinery) can we achieve our goals. This idea is all the more dangerous because no one can reasonably object to the rational use of techniques to achieve human purposes. Indeed, I am not disputing that the technique known as management may be the best way for modern business to conduct its affairs. We are technical creatures, and through our predilection for and our ability to create techniques we achieve high levels of clarity and efficiency. As I said earlier, language itself is a kind of technique—an invisible technology—and through it we achieve more than clarity and efficiency. We achieve humanity—or inhumanity. The question with language, as with any other technique or machine, is and always has been, Who is to be the master? Will we control it, or will it control us? The argument, in short, is not with technique. The argument is with the triumph of technique, with techniques that become sanctified and rule out the possibilities of other ones. Technique, like any other technology, tends to function independently of the system it serves. It becomes autonomous, in the manner of a robot that no longer obeys its master.

Second, management is an important example of how an "invisible technology" works subversively but powerfully to create a new way of doing things, a classic instance of the tail wagging the dog. It is entirely possible for business and other institutions to operate without a highly technicalized management structure, however hard for us to imagine. We have grown so accustomed to it that we are near to believing management is an aspect of the natural order of things, just as students and teachers have come to believe that education would be impossible without the structure of a college "course." And politicians believe they would be adrift without the assistance of public-opinion polling. When a method of doing things becomes so deeply associated with an institution that we no longer know which came first—the method or the institution—then it is difficult to change the institution or even to imagine alternative methods for achieving its purposes.

And so it is necessary to understand where our techniques come from and what they are good for; we must make them visible so that they may be restored to our sovereignty. In the next chapter, I hope to do this with the intricate and vast ensemble of techniques I call Scientism.

9

Scientism

On December 5, 1989, Daniel Goleman, covering the social-science beat for *The New York Times,* gave considerable space to some "recent research findings" that doubtless unsettled readers who hadn't been keeping informed about the work of our scientists of the mind: Goleman reported that psychological researchers have discovered that people fear death. This insight led them to formulate "a sweeping theory," to quote Goleman, "that gives the fear of death a central and often unsuspected role in psychological life." To whom death's role is unsuspected we were not told, but the theory is sufficiently rich to allow the hypothesis that all cultures (to quote Goleman again) "prescribe what people should do to lead a 'good' and 'meaningful' life and offer some hope of immortality, as in the the [sic] Christian afterlife or the Hindu notion of reincarnation into a better life." (The repetition of the word "the" in the sentence quoted above may have been a typographical error—or else perhaps an excited stammer in the face of such an astounding hypothesis.) As if this were not enough, Goleman also reported the same psychologists as having discovered that how one reacts to death depends on one's moral code, and that those who value open-mindedness are more tolerant of people whose values differ from theirs—which means that those who are open-minded tend to be open-minded, a fact that is not sufficiently appreciated, if known at all.

On September 11, 1990, Goleman revealed the results of new research which suggests that Asian-American students do well in school because they come from intact families that value advanced academic degrees. And on October 2, 1990, he reported that psychologists have discovered that children who are inept at social relations tend to be unpopular with other children.

I cite these reports from *The New York Times* because it is considered by many to be the "newspaper of public record" and may be assumed to be reporting the *best* of social science. It is possible, of course, that Goleman is a "mole," or an undercover agent, who is trying to reveal where our culture stands by ridiculing the trivialities of social science. But I doubt it. He seems to believe in social science, as so many in Technopoly do. That is, he believes that the study of human behavior, when conducted according to the rigorous principles established by the physical and biological sciences, will produce objective facts, testable theories, and profound understandings of the human condition. Perhaps even universal laws.

I have previously attributed the origins of this belief to the work of Auguste Comte, which is a defensible position but something of an oversimplification. In fact, the beginning formulations of a "science of man" are more precisely attributed to a school than to a man. The school, founded in 1794 in Paris, was called the Ecole Polytechnique (the same school that, as I mentioned earlier, quickly adopted the practice begun at Cambridge of assigning number grades to student work). The Ecole Polytechnique gathered for its teaching staff the best scientists, mathematicians, and engineers France had produced, and became famous for its enthusiasm for the methods of the natural sciences. Lavoisier and Ampère taught there, as did, later, Volta and Alexander von Humboldt. Their work in chemistry and physics helped to lay the foundation of modern science, and in that respect the Ecole Polytechnique is justly honored. But there were others associated with the school whose exuberance for the methods of the natural sciences led them to believe that there were no limits to the powers of the human mind, and in particular no limits to the power of scientific research. The most famous expression of what may be called "scientific hubris" appeared in Pierre-Simon de Laplace's *Essai philosophique sur les probabilités,* published in 1814. He wrote: "A mind that in a given instance knew all the forces by which nature is animated and the position of all the bodies of which it is composed, if it were vast enough to include all these data within his analysis, could embrace in one single formula the movements of the largest bodies of the universe and of the smallest atoms; nothing would be uncertain for him; the future and the past would be equally before his eyes."[1]

There is, of course, no scientist today who takes this view seriously, and there were few enough who did in the nineteenth century. But the spirit behind this scientific ideal inspired several men to believe that the reliable and predictable knowledge that could be obtained about stars and atoms could also be obtained about human behavior. Among the best known of these early "social scientists" were Claude-Henri de Saint-Simon, Prosper Enfantin, and, of course, Auguste Comte. They held in common two beliefs to which Technopoly is deeply indebted: that the natural sciences provide a method to unlock the secrets of both the human heart and the direction of social life; that society can be rationally and humanely reorganized according to principles that social science will uncover. It is with these men that the idea of "social engineering" begins and the seeds of Scientism are planted.

By Scientism, I mean three interrelated ideas that, taken together, stand as one of the pillars of Technopoly. Two of the three have just been cited. The first and indispensable idea is, as noted, that the methods of the natural sciences can be applied to the study of human behavior. This idea is the backbone of much of psychology and sociology as practiced at least in America, and largely accounts for the fact that social science, to quote F. A. Hayek, "has contributed scarcely anything to our understanding of social phenomena."[2]

The second idea is, as also noted, that social science generates specific principles which can be used to organize society on a rational and humane basis. This implies that technical means—mostly "invisible technologies" supervised by experts—can be designed to control human behavior and set it on the proper course.

The third idea is that faith in science can serve as a comprehensive belief system that gives meaning to life, as well as a sense of well-being, morality, and even immortality.

I wish here to show how these ideas spiral into each other, and how they give energy and form to Technopoly.

The term "science," as it is generally used today—referring to the work of those in the physical, chemical, and biological disciplines—was popularized in the early nineteenth century, with significant help from the formation of the British Association for the Advancement of Science in 1831 (although Murray's *New English Dictionary* gives 1867 as the earliest use of the term in its modern sense). By the early twentieth century, the term had been appropriated by others, and it has since become increasingly familiar as a description of what psychologists, sociologists, and even anthropologists do. It will come as no surprise that I claim this is a deceptive and confusing use of the term, in part because it blurs the distinction between processes and practices.

Using definitions proposed by the British philosopher Michael Oakeshott, we may say that "processes" refers to those events that occur in nature, such as the orbiting of planets or the melting of ice or the production of chlorophyll in a leaf. Such processes have nothing to do with human intelligence, are governed by immutable laws, and are, so to say, determined by the structure of nature. If one were so inclined, one might even say that processes are the creation of God. By "practices," on the other hand, Oakeshott means the creations of people—those events that result from human decisions and actions, such as writing or reading this book or forming a new government or conversing at dinner or falling in love. These events are a function of human intelligence interacting with environment, and although there is surely a measure of regularity in human affairs, such affairs are not determined by natural laws, immutable or otherwise. In other words, there is an irrevocable difference between a blink and a wink. A blink can be classified as a process; it has physiological causes which can be understood and explained within the context of established postulates and theories. But a wink must be classified as a practice, filled with personal and to some extent unknowable meanings and, in any case, quite impossible to explain or predict in terms of causal relations.

What we may call science, then, is the quest to find the immutable and universal laws that govern processes, presuming that there are cause-and-effect relations among these processes. It follows that the quest to understand human behavior and feeling can in no sense except the most trivial be called science. One can, of course, point to the fact that students of both natural law and human behavior often quantify their observations, and on this common ground classify them together. A fair analogy would be to argue that, since a housepainter and an artist both use paint, they are engaged in the same enterprise and to the same end.

The scientist uses mathematics to assist in uncovering and

describing the structure of nature. At best, sociologists (to take one example) use quantification merely to give some precision to their ideas. But there is nothing especially scientific in that. All sorts of people count things in order to achieve precision without claiming they are scientists. Bail bondsmen count the number of murders committed in their cities; judges count the number of divorce actions in their jurisdictions; business executives count the amount of money spent in their stores; and young children like to count their toes and fingers in order not to be vague about how many they have. Information produced by counting may sometimes be valuable in helping a person get an idea, or, even more so, in providing support for an idea. But the mere activity of counting does not make science.

Nor does observing things, though it is sometimes said that if one is empirical, one is scientific. To be empirical means to look at things before drawing conclusions. Everyone, therefore, is an empiricist, with the possible exception of paranoid schizophrenics. To be empirical also means to offer evidence that others can see as clearly as you. You may, for example, conclude that I like to write books, offering as evidence that I have written this one and several others besides. You may also offer as evidence a tape recording, which I can supply on request, on which I tell you that I like to write books. Such evidence may be said to be empirical, and your conclusion empirically based. But you are not therefore acting as a scientist. You are acting as a rational person, to which condition many people who are not scientists may make a just claim.

Scientists do strive to be empirical and where possible precise, but it is also basic to their enterprise that they maintain a high degree of objectivity, which means that they study things independently of what people think or do about them. The opinions people hold about the external world are, to scientists, always an obstacle to be overcome, and it is well known that the scientist's picture of the external world is quite different from what most people believe the world to be like. Moreover, in their quest for objectivity, scientists proceed on the assumption that the objects they study are indifferent to the fact that they are being studied. Heisenberg's uncertainty principle indicates that at subatomic levels particles do "know" they are being studied, at least in a special meaning of "knowing." An electron, for example, changes either its momentum or its position when it is being tracked—i.e., when it interacts with a photon—but the electron does not, in the usual sense of the word, "know" or "care" that the interaction is taking place. Nor do objects like leaves, apples, planets, kidneys, or bridges. This fact relieves the scientist of inquiring into their values and motivations and for this reason alone separates science from what is called social science, consigning the methodology of the latter (to quote Gunnar Myrdal) to the status of the "metaphysical and pseudo-objective."[3]

The status of social-science methods is further reduced by the fact that there are almost no experiments that will reveal a social-science theory to be false. Theories in social science disappear, apparently, because they are boring, not because they are refuted. But, as Karl Popper has demonstrated, science depends on the requirement that theories must be stated in a way that permits experiments to reveal that they are false. If a theory cannot be tested for its falsity, it is not a scientific theory—as, for example, Freud's theory of the Oedipus complex. Psychiatrists can provide many examples supporting the validity of the theory, but they have no answer to the question "What evidence would prove the theory false?" Believers in the God theory (sometimes called Creation Science) are silent on the question "What evidence would show that there is no God?"

I do not say, incidentally, that the Oedipus complex and God do not exist. Nor do I say that to believe in them is harmful—far from it. I say only that, there being no tests that could, in principle, show them to be false, they fall outside the purview of science, as do almost all theories that make up the content of "social science."

I shall say in a few moments what I believe social science to be, as well as why Technopoly wishes to link it to the scientific enterprise. Here, I should like to give an example of social science to amplify the reasons why it is misleading to call it science.

A piece of work that is greatly admired as social science, at least from a technical if not an ethical point of view, is the set of experiments (so called) supervised by Stanley Milgram, the account of which was published under the title *Obedience to Authority.* In this notorious study, Milgram sought to entice people to give electric shocks to "innocent victims" who were in fact conspirators in the experiment and did not actually receive the shocks. Nonetheless, most of Milgram's subjects *believed* that the victims were receiving the shocks, and many of them, under psychological pressure, gave shocks that, had they been real, might have killed the victims. Milgram took great care in designing the environment in which all this took place, and his book is filled with statistics that indicate how many did or did not do what the experimenters told them to do. Somewhere in the neighborhood of 65 percent of his subjects were rather more compliant than would have been good for the health of their victims. Milgram drew the following conclusion from his research: In the face of what they construe to be legitimate authority, most people will do what they are told. Or, to put it another way, the social context in which people find themselves will be a controlling factor in how they behave.

Now, in the first place, this conclusion is merely a commonplace of human experience, known by just about everyone from Maimonides to your aunt and uncle. The exceptions seem to be American psychiatrists. Before he conducted his experiment, Milgram sent a questionnaire to a large group of psychiatrists from whom he solicited opinions as to how many subjects would be likely to continue giving electric shocks when ordered to do so. The psychiatrists thought the number would be very much smaller than it actually was, basing their estimates on their knowledge of human behavior (which only recently has admitted the idea that people fear death). I do not mean to imply that real scientists never produce commonplaces, but only that it is rare, and never a cause for excitement. On the other hand, commonplace conclusions are almost always a characteristic of social research pretending to be science.

In the second place, Milgram's study was not empirical in the strict sense, since it was not based on observations of people in natural life situations. I assume that no one is especially interested in how people behave in a laboratory at Yale or any other place; what matters is how people behave in situations where their behavior makes a difference to their lives. But any conclusions that can be drawn from Milgram's study must specify that they apply only to people in laboratories under the conditions Milgram arranged. And even if we assume a correspondence between laboratory behavior and more lifelike situations, no predictions can be made about *what* lifelike situations these might be. Nor can any serious claim be made that there is a causal relationship between the acceptance of legitimate authority and doing what you are told. In fact, Milgram himself shows us that there is not, since 35 percent of his subjects told the "authority figure" to bug off. Moreover, Milgram had no idea *why* some people did and some people did not tell him to bug off. For myself, I feel quite sure that if each of Milgram's subjects had been required to read Hannah Arendt's *Eichmann in Jerusalem* before showing up at the laboratory, his numbers would have been quite different.

But let us suppose that I am wrong about that, and let us further suppose that Milgram had found that 100 percent of his subjects did what they were told, with or without Hannah Arendt. And now let us suppose that I tell you a story of a group of people who in some real situation refused to comply with the orders of a legitimate authority—let us say, the Danes who in the face of Nazi occupation helped nine thousand Jews escape to Sweden. Would you say to me that this cannot be so because Milgram's study proves otherwise? Or would you say that this overturns Milgram's work? Perhaps you would say that the Danish response is not relevant, since the Danes did not regard the Nazi occupation as constituting legitimate authority. But then, how would we explain the cooperative response to Nazi authority of the French, the Poles, and the Lithuanians? I think you would say none of these things, because Milgram's experiment does not confirm or falsify any theory that might be said to postulate a law of human nature. His study—which, incidentally, I find both fascinating and terrifying—is not science. It is something else entirely.

Which leads me to say what sort of work I think Milgram was engaged in—and what sort of work those who study human behavior and situations are engaged in. I will start by making reference to a famous correspondence between Sigmund Freud and Albert Einstein. Freud once sent a copy of one of his books to Einstein, asking for his evaluation of it. Einstein replied that he thought the book exemplary but was not qualified to judge its scientific merit. To which Freud replied somewhat testily that, if Einstein could say nothing of its scientific merit, he, Freud, could not imagine how the book could be judged exemplary: it was science or it was nothing. Well, of course, Freud was wrong. His work *is* exemplary—indeed, monumental—but scarcely anyone believes today that Freud was doing science, any more than educated people believe that Marx was doing science, or Max Weber or Lewis Mumford or Bruno Bettelheim or Carl Jung or Margaret Mead or Arnold Toynbee. What these people were doing—and Stanley Milgram was doing—is documenting the behavior and feelings of people as they confront problems posed by their culture. Their work is a form of storytelling. Science itself is, of course, a form of storytelling too, but its assumptions and procedures are so different from those of social research that it is extremely misleading to give the same name to each. In fact, the stories of social researchers are much closer in structure and purpose to what is called imaginative literature; that is to say, both a social researcher and a novelist give unique interpretations to a set of human events and support their interpretations with examples in various forms. Their interpretations cannot be proved or disproved but will draw their appeal from the power of their language, the depth of their explanations, the relevance of their examples, and the credibility of their themes. And all of this has, in both cases, an identifiable moral purpose. The words "true" and "false" do not apply here in the sense that they are used in mathematics or science. For there is nothing universally and irrevocably true or false about these interpretations. There are no critical tests to confirm or falsify them. There are no natural laws from which they are derived. They are bound by time, by situation, and above all by the cultural prejudices of the researcher or writer.

A novelist—for example, D. H. Lawrence—tells a story about the sexual life of a woman—Lady Chatterley—and from it we may learn things about the secrets of some people, and wonder if Lady Chatterley's secrets are not more common than we had thought. Lawrence did not claim to be a scientist, but he looked carefully and deeply at the people he knew and concluded that there is more hypocrisy in heaven and earth than is dreamt of in some of our philosophies. Alfred Kinsey was also interested in the sexual lives of women, and so he and his assistants interviewed thousands of them in an effort to find out what they believed their sexual conduct was like. Each woman told her story, although it was a story carefully structured by Kinsey's questions. Some of them told everything they were permitted to tell, some only a little, and some probably lied. But when all their tales were put together, a collective story emerged about a certain time and place. It was a story more abstract than D. H. Lawrence's, largely told in the language of statistics and, of course, without much psychological insight. But it was a story nonetheless. One might call it a tribal tale of one thousand and one nights, told by a thousand and one women, and its theme was not much different from Lawrence's—namely, that the sexual life of some women is a lot stranger and more active than some other stories, particularly Freud's, had led us to believe.

I do not say that there is no difference between Lawrence and Kinsey. Lawrence unfolds his story in a language structure called a narrative. Kinsey's language structure is called exposition. These forms are certainly different, although not so much as we might suppose. It has been remarked about the brothers Henry and William James that Henry was the novelist who wrote like a psychologist, and William the psychologist who wrote like a novelist. Certainly, in my meaning of the word "story," exposition is as capable of unfolding one as is narrative. Of course, Lawrence's story is controlled entirely by the limits of his own imagination, and he is not obliged to consult any social facts other than those he believed he knew. His story is pure personal perception, and that is why we call it fiction. Kinsey's story comes from the mouths of others, and he is limited by what they answered when he asked his questions. Kinsey's story, therefore, we may call a documentary. But, like all stories, it is infused with moral prejudice and sociological theory. It is Kinsey who made up the questions, and chose who would be interviewed, the circumstances of the interview, and how the answers would be interpreted. All of this gives shape and point to his story. Indeed, we may assume that Kinsey, like Lawrence, knew from the outset what the theme of his story would be. Otherwise, he probably wouldn't have cared to tell it.

Both the novelist and the social researcher construct their stories by the use of archetypes and metaphors. Cervantes, for example, gave us the enduring archetype of the incurable dreamer and idealist in Don Quixote. The social historian Marx gave us the archetype of the ruthless and conspiring, though nameless, capitalist. Flaubert gave us the repressed bourgeois romantic in Emma Bovary. And Margaret Mead gave us the carefree, guiltless Samoan adolescent. Kafka gave us the alienated urbanite driven to self-loathing. And Max Weber gave us hardworking men driven by a mythology he called the Protestant Ethic. Dostoevsky gave us the egomaniac redeemed by love and religious fervor. And B. F. Skinner gave us the automaton redeemed by a benign technology.

I think it justifiable to say that, in the nineteenth century, novelists provided us with most of the powerful metaphors and images of our culture. In the twentieth century, such metaphors and images have come largely from the pens of social historians and researchers. Think of John Dewey, William James, Erik Erikson, Alfred Kinsey, Thorstein Veblen, Margaret Mead, Lewis Mumford, B. F. Skinner, Carl Rogers, Marshall McLuhan, Barbara Tuchman, Noam Chomsky, Robert Coles, even Stanley Milgram, and you must acknowledge that our ideas of what we are like and what kind of country we live in come from their stories to a far greater extent than from the stories of our most renowned novelists.

I do not mean, incidentally, that the metaphors of social research are created in the same way as those of novels and plays. The writer of fiction creates metaphors by an elaborate and concrete detailing of the actions and feelings of particular human beings. Sociology is background; individual psychology is the focus. The researcher tends to do it the other way around. The focus is on a wider field, and the individual life is seen in silhouette, by inference and suggestion. Also, the novelist proceeds by showing. The researcher, using abstract social facts, proceeds by reason, by logic, by argument. That is why fiction is apt to be more entertaining. Whereas Oscar Wilde or Evelyn Waugh *shows* us the idle and conspicuously consuming rich, Thorstein Veblen *argues* them into existence. In the character of Sammy Glick, Budd Schulberg *shows* us the narcissist whose origins Christopher Lasch has tried to *explain* through sociological analysis. So there are differences among storytellers, and most of the time our novelists are more pleasurable to read. But the stories told by our social researchers are at least as compelling and, in our own times, apparently more credible.

Why do such social researchers tell their stories? Essentially for didactic and moralistic purposes. These men and women tell their stories for the same reason the Buddha, Confucius, Hillel, and Jesus told their stories (and for the same reason D. H. Lawrence told his). It is true, of course, that social researchers rarely base their claims to knowledge on the indisputability of sacred texts, and even less so on revelation. But we must not be dazzled or deluded by differences in method between preachers and scholars. Without meaning to be blasphemous, I would say that Jesus was as keen a sociologist as Veblen. Indeed, Jesus' remark about rich men, camels, and the eye of a needle is as good a summary of Veblen's *Theory of the Leisure Class* as it is possible to make. As social researchers, Jesus and Veblen differed in that Veblen was more garrulous.[4]

Unlike science, social research never discovers anything. It only rediscovers what people once were told and need to be told again. If, indeed, the price of civilization is repressed sexuality, it was not Sigmund Freud who discovered it. If the consciousness of people is formed by their material circumstances, it was not Marx who discovered it. If the medium is the message, it was not McLuhan who discovered it. They have merely retold ancient stories in a modern style. And these stories will be told anew decades and centuries from now, with, I imagine, less effect. For it would seem that Technopoly does not want these kinds of stories but facts—hard facts, scientific facts. We might even say that in Technopoly precise knowledge is preferred to truthful knowledge but that in any case Technopoly wishes to solve, once and for all, the dilemma of subjectivity. In a culture in which the machine, with its impersonal and endlessly repeatable operations, is a controlling metaphor and considered to be the instrument of progress, subjectivity becomes profoundly unacceptable. Diversity, complexity, and ambiguity of human judgment are enemies of technique. They mock statistics and polls and standardized tests and bureaucracies. In Technopoly, it is not enough for social research to rediscover ancient truths or to comment on and criticize the moral behavior of people. In Technopoly, it is an insult to call someone a "moralizer." Nor is it sufficient for social research to put forward metaphors, images, and ideas that can help people live with some measure of understanding and dignity. Such a program lacks the aura of certain knowledge that only science can provide. It becomes necessary, then, to transform psychology, sociology, and anthropology into "sciences," in which humanity itself becomes an object, much like plants, planets, or ice cubes.

That is why the commonplaces that people fear death and that children who come from stable families valuing scholarship will do well in school must be announced as "discoveries" of scientific enterprise. In this way, social researchers can see themselves, and can be seen, as scientists, researchers without bias or values, unburdened by mere opinion. In this way, social policies can be claimed to rest on objectively determined facts. In Technopoly, it is not enough to argue that the segregation of blacks and whites in schools is immoral, and it is useless to offer *Black Boy* or *Invisible Man* or *The Fire Next Time* as proof. The courts must be shown that standardized academic and psychological tests reveal that blacks do less well than whites and feel demeaned when segregation exists. In Technopoly, it is not enough to say it is immoral and degrading to allow people to be homeless. You cannot get anywhere by asking a judge, a politician, or a bureaucrat to read *Les Misérables* or *Nana* or, indeed, the New Testament. You must show that statistics have produced data revealing the homeless to be unhappy and to be a drain on the economy. Neither Dostoevsky nor Freud, Dickens nor Weber, Twain nor Marx, is now a dispenser of legitimate knowledge. They are interesting; they are "worth reading"; they are artifacts of our past. But as for "truth," we must turn to "science." Which brings me to the crux of what I mean by Scientism, and why it has emerged in Technopoly.

I have tried to show that science, social research, and the kind of work we call imaginative literature are three quite different kinds of enterprise. In the end, *all* of them are forms of storytelling—human attempts to account for our experience in coherent ways. But they have different aims, ask different questions, follow different procedures, and give different meanings to "truth." In most of these respects, social research has little in common with science, and much in common with other forms of imaginative literature. Yet social "scientists" have consistently sought to identify themselves, and in more than name, with physicists, chemists, biologists, and others who inquire into the lawful regularities of the natural world. Why students of the *human* condition should do this is not hard to explain. The great successes of modern times—indeed, perhaps the only successes—have come in medicine, pharmacology, biochemistry, astrophysics, and all the feats of mechanical, biological, and electronic engineering made possible by the consistent application of the aims, assumptions, and procedures of natural science. These successes have attached to the name of science an awesome measure of authority, and to those who claim the title "scientist" a similar measure of respect and prestige. Beyond that lies the nineteenth-century hope that the assumptions and procedures of natural science might be applied without modification to the social world, to the same end of increased predictability and control, and with the same kind of engineering success. This hope has proved both misguided and illusory. But the illusion is a powerful one, and, given the psychological, social, and material benefits that attach to the label "scientist," it is not hard to see why social researchers should find it hard to give it up.

It is less easy to see why the rest of us have so willingly, even eagerly, cooperated in perpetuating the same illusion. In part, the explanation lies in a profound misunderstanding of the aims of natural and of social studies, and of the differences between the physical and social worlds. But there is more to it than that. When the new technologies and techniques and spirit of men like Galileo, Newton, and Bacon laid the foundations of natural science, they also discredited the authority of earlier accounts of the physical world, as found, for example, in the great tale of Genesis. By calling into question the truth of such accounts in one realm, science undermined the whole edifice of belief in sacred stories and ultimately swept away with it the source to which most humans had looked for *moral* authority. It is not too much to say, I think, that the desacralized world has been searching for an alternative source of moral authority ever since. So far as I know, no responsible natural scientist, either of the Renaissance or of recent times, has claimed that the procedures of natural science or its discoveries can tell us what we *ought* to do—whether some way of dealing with our fellow humans is good or evil, right or wrong. Indeed, the very principles of natural science, with its requirement of an objective stance toward what is studied, compel the natural scientist to abjure *in his or her role as a scientist* such moral judgments or claims. When natural scientists speak out on moral questions, on what is good or evil to do, they speak as the rest of us—as concerned citizens on a threatened planet, as rational women and men, as people of conscience who must struggle no less than you must, or I, to answer for themselves where the ultimate authority for their moral judgment lies. It is the world of desperate listeners, longing for a more powerful moral authority, that begs the natural scientist to say it is *the science* that speaks, not the woman or man. But the scientist cannot with honor consent.

Our social "scientists" have from the beginning been less tender of conscience, or less rigorous in their views of science, or perhaps just more confused about the questions their procedures can answer and those they cannot. In any case, they have not been squeamish about imputing to their "discoveries" and the rigor of their procedures the power to direct us in how we ought rightly to behave. That is why social "scientists" are so often to be found on our television screens, and on our best-seller lists, and in the "self-help" sections of airport bookstands: not because they can tell us how some humans sometimes behave but because they purport to tell us how we *should;* not because they speak to us as fellow humans who have lived longer, or experienced more of human suffering, or thought more deeply and reasoned more carefully about some set of problems, but because they consent to maintain the illusion that it is their data, their procedures, their science, and not themselves, that speak. We welcome them gladly, and the claim explicitly made or implied, because we need so desperately to find some source outside the frail and shaky judgments of mortals like ourselves to authorize our moral decisions and behavior. And outside of the authority of brute force, which can scarcely be called moral, we seem to have little left but the authority of procedures.

This, then, is what I mean by Scientism. It is not merely the misapplication of techniques such as quantification to questions where numbers have nothing to say; not merely the confusion of the material and social realms of human experience; not merely the claim of social researchers to be applying the aims and procedures of natural science to the human world. Scientism is all of these, but something profoundly more. It is the desperate hope, and wish, and ultimately the illusory belief that some standardized set of procedures called "science" can provide us with an unimpeachable source of moral authority, a suprahuman basis for answers to questions like "What is life, and when, and why?" "Why is death, and suffering?" "What is right and wrong to do?" "What are good and evil ends?" "How ought we to think and feel and behave?" It is Scientism on a personal level when one says, as President Reagan did, that he personally believes that abortion is wrong but we must leave it to science to tell us when a fetus enters life. It is Scientism on a cultural level when no scientist rises to demur, when no newspaper prints a rebuttal on its "science" pages, when everyone cooperates, willfully or through ignorance, in the perpetuation of such an illusion. Science can tell us when a heart begins to beat, or movement begins, or what are the statistics on the survival of neonates of different gestational ages outside the womb. But science has no more authority than you do or I do to establish such criteria as the "true" definition of "life" or of human state or of personhood. Social research can tell us how some people behave in the presence of what they believe to be legitimate authority. But it cannot tell us when authority is "legitimate" and when not, or how we must decide, or when it may be right or wrong to obey. To ask of science, or expect of science, or accept unchallenged from science the answers to such questions is Scientism. And it is Technopoly's grand illusion.

Toward the end of his life, Sigmund Freud debated with himself what he called *The Future of an Illusion.* The illusion he referred to was the belief in a supranatural and suprahuman source of being, knowledge, and moral authority: the belief in God. The question Freud debated was not whether God exists, but whether humankind could survive without the illusion of God—or, rather, whether humankind would fare better psychologically, culturally, and morally without that illusion than with it. Freud states his own doubts (expressed through the device of an alter ego with whom he debates) in the strongest possible voice, but in the end it is the voice of Freud's reason (or faith in reason) that "wins": humankind may or may not fare better, but it must do without the illusion of God. Freud did not see that, even as he wrote, his own work was lending substance to another illusion: the illusion of a future in which the procedures of natural and social science would ultimately reveal the "real" truth of human behavior and provide, through the agency of objectively neutral scientists, an empirical source of moral authority. Had he foreseen the peculiar transformation that the image of ultimate authority would take in our own time—from an old man in a long white beard to young men and women in long white coats—Freud might have changed the question that was the focus of his inquiry. He could not. And so I will change it here, not to provide an answer, but in the hope of stirring renewed debate: as among the illusion of God, the illusion of Scientism, and no illusion or hope at all for an ultimate source of moral authority, which is most likely to serve the human interest, and which to prove most deadly, in the Age of Technopoly?

10

The Great Symbol Drain

It is possible that, some day soon, an advertising man who must create a television commercial for a new California Chardonnay will have the following inspiration: Jesus is standing alone in a desert oasis. A gentle breeze flutters the leaves of the stately palms behind him. Soft Mideastern music caresses the air. Jesus holds in his hand a bottle of wine at which he gazes adoringly. Turning toward the camera, he says, "When I transformed water into wine at Cana, *this* is what I had in mind. Try it today. You'll become a believer."

If you think such a commercial is not possible in your lifetime, then consider this: As I write, there is an oft-seen commercial for Hebrew National frankfurters. It features a dapper-looking Uncle Sam in his traditional red, white, and blue outfit. While Uncle Sam assumes appropriate facial expressions, a voice-over describes the delicious and healthful frankfurters produced by Hebrew National. Toward the end of the commercial, the voice stresses that Hebrew National frankfurters surpass federal standards for such products. Why? Because, the voice says as the camera shifts our point of view upward toward heaven, "We have to answer to a Higher Authority."

I will leave it to the reader to decide which is more incredible—Jesus being used to sell wine or God being used to sell frankfurters. Whichever you decide, you must keep in mind that neither the hypothetical commercial nor the real one is an example of blasphemy. They are much worse than that. Blasphemy is, after all, among the highest tributes that can be paid to the power of a symbol. The blasphemer takes symbols as seriously as the idolater, which is why the President of the United States (circa 1991) wishes to punish, through a constitutional amendment, desecrators of the American flag.

What we are talking about here is not blasphemy but trivialization, against which there can be no laws. In Technopoly, the trivialization of significant cultural symbols is largely conducted by commercial enterprise. This occurs not because corporate America is greedy but because the adoration of technology pre-empts the adoration of anything else. Symbols that draw their meaning from traditional religious or national contexts must therefore be made impotent as quickly as possible—that is, drained of sacred or even serious connotations. The elevation of one god requires the demotion of another. "Thou shalt have no other gods before me" applies as well to a technological divinity as any other.

There are two intertwined reasons that make it possible to trivialize traditional symbols. The first, as neatly expressed by the social critic Jay Rosen, is that, although symbols, especially images, are endlessly repeatable, they are not inexhaustible. Second, the more frequently a significant symbol is used, the less potent is its meaning. This is a point stressed in Daniel Boorstin's classic book *The Image,* published thirty years ago.[1] In it, Boorstin describes the beginnings, in the mid-nineteenth century, of a "graphics revolution" that allowed the easy reproduction of visual images, thus providing the masses with continuous access to the symbols and icons of their culture. Through prints, lithographs, photographs, and, later, movies and television, religious and national symbols became commonplaces, breeding indifference if not necessarily contempt. As if to answer those who believe that the emotional impact of a sacred image is always and ever the same, Boorstin reminds us that prior to the graphics revolution most people saw relatively few images. Paintings of Jesus or the Madonna, for example, would have been seen rarely outside churches. Paintings of great national leaders could be seen only in the homes of the wealthy or in government buildings. There were images to be seen in books, but books were expensive and spent most of their time on shelves. Images were not a conspicuous part of the environment, and their scarcity contributed toward their special power. When the scale of accessibility was altered, Boorstin argues, the experience of encountering an image necessarily changed; that is to say, it diminished in importance. One picture, we are told, is worth a thousand words. But a thousand pictures, especially if they are of the same object, may not be worth anything at all.

What Boorstin and Rosen direct our attention to is a common enough psychological principle. You may demonstrate this for yourself (if you have not at some time already done so) by saying any word, even a significant one, over and over again. Sooner than you expect, you will find that the word has been transformed into a meaningless sound, as repetition drains it of its symbolic value. Any male who has served in, let us say, the United States Army or spent time in a college dormitory has had this experience with what are called obscene words, especially the notorious four-letter word which I am loath to reproduce here. Words that you have been taught not to use and that normally evoke an embarrassed or disconcerted response, when used too often, are stripped of their power to shock, to embarrass, to call attention to a special frame of mind. They become only sounds, not symbols.

Moreover, the journey to meaninglessness of symbols is a function not only of the frequency with which they are invoked but of the indiscriminate contexts in which they are used. An obscenity, for example, can do its work best when it is reserved for situations that call forth anger, disgust, or hatred. When it is used as an adjective for every third noun in a sentence, irrespective of the emotional context, it is deprived of its magical effects and, indeed, of its entire point. This is what happens when Abraham Lincoln's image, or George Washington's, is used to announce linen sales on Presidents' Day, or Martin Luther King's birthday celebration is taken as an occasion for furniture discounts. It is what happens when Uncle Sam, God, or Jesus is employed as an agent of the profane world for an essentially trivial purpose.

An argument is sometimes made that the promiscuous use of sacred or serious symbols by corporate America is a form of healthy irreverence. Irreverence, after all, is an antidote to excessive or artificial piety, and is especially necessary when piety is used as a political weapon. One might say that irreverence, not blasphemy, is the ultimate answer to idolatry, which is why most cultures have established means by which irreverence may be expressed—in the theater, in jokes, in song, in political rhetoric, even in holidays. The Jews, for example, use Purim as one day of the year on which they may turn a laughing face on piety itself.

But there is nothing in the commercial exploitation of traditional symbols that suggests an excess of piety is itself a vice. Business is too serious a business for that, and in any case has no objection to piety, as long as it is directed toward the idea of consumption, which is never treated as a laughing matter. In using Uncle Sam or the flag or the American Eagle or images of presidents, in employing such names as Liberty Insurance, Freedom Transmission Repair, and Lincoln Savings and Loan, business does not offer us examples of irreverence. It is merely declaring the irrelevance, in Technopoly, of distinguishing between the sacred and the profane.

I am not here making a standard-brand critique of the excesses of capitalism. It is entirely possible to have a market economy that respects the seriousness of words and icons, and which disallows their use in trivial or silly contexts. In fact, during the period of greatest industrial growth in America—from roughly 1830 to the end of the nineteenth century—advertising did not play a major role in the economy, and such advertising as existed used straightforward language, without recourse to the exploitation of important cultural symbols. There was no such thing as an "advertising industry" until the early twentieth century, the ground being prepared for it by the Postal Act of March 3, 1879, which gave magazines low-cost mailing privileges. As a consequence, magazines emerged as the best available conduits for national advertising, and merchants used the opportunity to make the names of their companies important symbols of commercial excellence. When George Eastman invented the portable camera in 1888, he spent $25,000 advertising it in magazines. By 1895, "Kodak" and "camera" were synonymous, as to some extent they still are. Companies like Royal Baking Powder, Baker's Chocolate, Ivory Soap, and Gillette moved into a national market by advertising their products in magazines. Even magazines moved into a national market by advertising themselves in magazines, the most conspicuous example being *Ladies' Home Journal*, whose publisher, Cyrus H. K. Curtis, spent half a million dollars between 1883 and 1888 advertising his magazine in other magazines. By 1909, *Ladies' Home Journal* had a circulation of more than a million readers.

Curtis' enthusiasm for advertising notwithstanding, the most significant figure in mating advertising to the magazine was Frank Munsey, who upon his death in 1925 was eulogized by William Allen White with the following words: "Frank Munsey contributed to the journalism of his day the talent of a meat packer, the morals of a money changer and the manners of an undertaker. He and his kind have about succeeded in transforming a once-noble profession into an 8% security. May he rest in trust." What was the sin of the malevolent Munsey? Simply, he made two discoveries. First, a large circulation could be achieved by selling a magazine for much less than it cost to produce it; second, huge profits could be made from the high volume of advertising that a large circulation would attract. In October 1893, Munsey took out an ad in the New York *Sun* announcing that *Munsey's Magazine* was cutting its price from 25 cents to 10 cents, and reducing a year's subscription from $3 to $1. The first 10-cent issue claimed a circulation of forty thousand; within four months, the circulation rose to two hundred thousand; two months later, it was five hundred thousand.

Munsey cannot, however, be blamed for another discovery, which for convenience's sake we may attribute to Procter and Gamble: that advertising is most effective when it is irrational. By irrational, I do not, of course, mean crazy. I mean that products could best be sold by exploiting the magical and even poetical powers of language and pictures. In 1892, Procter and Gamble invited the public to submit rhymes to advertise Ivory Soap. Four years later, H-O employed, for the first time, a picture of a baby in a high chair, the bowl of H-O cereal before him, his spoon in hand, his face ecstatic. By the turn of the century, advertisers no longer assumed that reason was the best instrument for the communication of commercial products and ideas. Advertising became one part depth psychology, one part aesthetic theory. In the process, a fundamental principle of capitalist ideology was rejected: namely, that the producer and consumer were engaged in a rational enterprise in which consumers made choices on the basis of a careful consideration of the quality of a product and their own self-interest. This, at least, is what Adam Smith had in mind. But today, the television commercial, for example, is rarely about the character of the products. It is about the character of the consumers of products. Images of movie stars and famous athletes, of serene lakes and macho fishing trips, of elegant dinners and romantic interludes, of happy families packing their station wagons for a picnic in the country—these tell nothing about the products being sold. But they tell everything about the fears, fancies, and dreams of those who might buy them. What the advertiser needs to know is not what is right about the product but what is wrong about the buyer. And so the balance of business expenditures shifts from product research to market research, which means orienting business away from making products of value and toward making consumers feel valuable. The business of business becomes pseudo-therapy; the consumer, a patient reassured by psychodramas.

What this means is that somewhere near the core of Technopoly is a vast industry with license to use all available symbols to further the interests of commerce, by devouring the psyches of consumers. Although estimates vary, a conservative guess is that the average American will have seen close to two million television commercials by age sixty-five. If we add to this the number of radio commercials, newspaper and magazine ads, and billboards, the extent of symbol overload and therefore symbol drain is unprecedented in human history. Of course, not all the images and words used have been cannibalized from serious or sacred contexts, and one must admit that as things stand at the moment it is quite unthinkable for the image of Jesus to be used to sell wine. At least not a chardonnay. On the other hand, his birthday is used as an occasion for commerce to exhaust nearly the entire repertoire of Christian symbology. The constraints are so few that we may call this a form of cultural rape, sanctioned by an ideology that gives boundless supremacy to technological progress and is indifferent to the unraveling of tradition.

In putting it this way, I mean to say that mass advertising is not the cause of the great symbol drain. Such cultural abuse could not have occurred without technologies to make it possible and a world-view to make it desirable. In the institutional form it has taken in the United States, advertising is a symptom of a world-view that sees tradition as an obstacle to its claims. There can, of course, be no functioning sense of tradition without a measure of respect for symbols. Tradition is, in fact, nothing but the acknowledgment of the authority of symbols and the relevance of the narratives that gave birth to them. With the erosion of symbols there follows a loss of narrative, which is one of the most debilitating consequences of Technopoly's power.

We may take as an example the field of education. In Technopoly, we improve the education of our youth by improving what are called "learning technologies." At the moment, it is considered necessary to introduce computers to the classroom, as it once was thought necessary to bring closed-circuit television and film to the classroom. To the question "Why should we do this?" the answer is: "To make learning more efficient and more interesting." Such an answer is considered entirely adequate, since in Technopoly efficiency and interest need no justification. It is, therefore, usually not noticed that this answer does not address the question "What is learning for?" "Efficiency and interest" is a technical answer, an answer about means, not ends; and it offers no pathway to a consideration of educational philosophy. Indeed, it blocks the way to such a consideration by beginning with the question of how we should proceed rather than with the question of why. It is probably not necessary to say that, by definition, there can be no education philosophy that does not address what learning is for. Confucius, Plato, Quintilian, Cicero, Comenius, Erasmus, Locke, Rousseau, Jefferson, Russell, Montessori, Whitehead, and Dewey—each believed that there was some transcendent political, spiritual, or social idea that must be advanced through education. Confucius advocated teaching "the Way" because in tradition he saw the best hope for social order. As our first systematic fascist, Plato wished education to produce philosopher kings. Cicero argued that education must free the student from the tyranny of the present. Jefferson thought the purpose of education is to teach the young how to protect their liberties. Rousseau wished education to free the young from the unnatural constraints of a wicked and arbitrary social order. And among John Dewey's aims was to help the student function without certainty in a world of constant change and puzzling ambiguities.

Only in knowing something of the reasons why they advocated education can we make sense of the means they suggest. But to understand their reasons we must also understand the narratives that governed their view of the world. By narrative, I mean a story of human history that gives meaning to the past, explains the present, and provides guidance for the future. It is a story whose principles help a culture to organize its institutions, to develop ideals, and to find authority for its actions. At the risk of repetition, I must point out again that the source of the world's greatest narratives has been religion, as found, for example, in Genesis or the Bhagavad-Gita or the Koran. There are those who believe—as did the great historian Arnold Toynbee—that without a comprehensive religious narrative at its center a culture must decline. Perhaps. There are, after all, other sources—mythology, politics, philosophy, and science, for example—but it is certain that no culture can flourish without narratives of transcendent origin and power.

This does not mean that the mere existence of such a narrative ensures a culture's stability and strength. There are destructive narratives. A narrative provides meaning, not necessarily survival—as, for example, the story provided by Adolf Hitler to the German nation in the 1930s. Drawing on sources in Teutonic mythology and resurrecting ancient and primitive symbolism, Hitler wove a tale of Aryan supremacy that lifted German spirits, gave point to their labors, eased their distress, and provided explicit ideals. The story glorified the past, elucidated the present, and foretold the future, which was to last a thousand years. The Third Reich lasted exactly twelve years.

It is not to my point to dwell on the reasons why the story of Aryan supremacy could not endure. The point is that cultures must have narratives and will find them where they will, even if they lead to catastrophe. The alternative is to live without meaning, the ultimate negation of life itself. It is also to the point to say that each narrative is given its form and its emotional texture through a cluster of symbols that call for respect and allegiance, even devotion. The United States Constitution, for example, is only in part a legal document, and, I should add, a small part. Democratic nations—England, for one—do not require a written constitution to ensure legal order and the protection of liberties. The importance of the American Constitution is largely in its function as a symbol of the story of our origins. It is our political equivalent of Genesis. To mock it, to ignore it, to circumvent it is to declare the irrelevance of the story of the United States as a moral light unto the world. In like fashion, the Statue of Liberty is the key symbol of the story of America as the natural home of the teeming masses, from anywhere, yearning to be free. There are, of course, several reasons why such stories lose their force. This book is, in fact, an attempt to describe one of them—i.e., how the growth of Technopoly has overwhelmed earlier, more meaningful stories. But in all cases, the trivialization of the symbols that express, support, and dramatize the story will accompany the decline. Symbol drain is both a symptom and a cause of a loss of narrative.

The educators I referred to above based their philosophies on narratives rich in symbols which they respected and which they understood to be integral to the stories they wanted education to reveal. It is, therefore, time to ask, What story does American education wish to tell now? In a growing Technopoly, what do we believe education is for? The answers are discouraging, and one of them can be inferred from any television commercial urging the young to stay in school. The commercial will either imply or state explicitly that education will help the persevering student to get a good job. And that's it. Well, not quite. There is also the idea that we educate ourselves to compete with the Japanese or the Germans in an economic struggle to be number one. Neither of these purposes is, to say the least, grand or inspiring. The story each suggests is that the United States is not a culture but merely an economy, which is the last refuge of an exhausted philosophy of education. This belief, I might add, is precisely reflected in the President's Commission Report, *A Nation at Risk*, where you will find a definitive expression of the idea that education is an instrument of economic policy and of very little else.

We may get a sense of the desperation of the educator's search for a more gripping story by using the "television commercial test." Try to imagine what sort of appeals might be effectively made on a TV commercial to persuade parents to support schools. (Let us, to be fair, sidestep appeals that might be made directly to students themselves, since the youth of any era are disinclined to think schooling a good idea, whatever the reasons advanced for it. See the "Seven Ages of Man" passage in *As You Like It*.)

Can you imagine, for example, what such a commercial would be like if Jefferson or John Dewey prepared it? "Your children are citizens in a democratic society," the commercial might say. "Their education will teach them how to be valuable citizens by refining their capacity for reasoned thought and strengthening their will to protect their liberties. As for their jobs and professions, that will be considered only at a 'late and convenient hour' " (to quote John Stuart Mill, who would be pleased to associate himself with Jefferson's or Dewey's purpose). Is there anyone today who would find this a compelling motivation? Some, perhaps, but hardly enough to use it as the basis of a national program. John Locke's commercial would, I imagine, be even less appealing. "Your children must stay in school," he might say, "because there they will learn to make their bodies slaves of their minds. They will learn to control their impulses, and how to find satisfaction and even excitement in the life of the mind. Unless they accomplish this, they can be neither civilized nor literate." How many would applaud this mission? Indeed, whom could we use to speak such words—Barbara Bush? Lee Iacocca? Donald Trump? Even the estimable Dr. Bill Cosby would hardly be convincing. The guffaws would resound from Maine to California.

In recent years, a valiant attempt has been made by some—for example, E. D. Hirsch, Jr.—to provide a comprehensive purpose to education. In his book *Cultural Literacy*, Hirsch defines literacy as the capacity to understand and use the words, dates, aphorisms, and names that form the basis of communication among the educated in our culture. Toward this end, he and some of his colleagues compiled a list that contains, according to them, the references essential to a culturally literate person in America. The first edition of the book (1987) included Norman Mailer but not Philip Roth, Bernard Malamud, Arthur Miller, or Tennessee Williams. It included Ginger Rogers but not Richard Rodgers, Carl Rogers, or Buck Rogers, let alone Fred Rogers. The second greatest home-run hitter of all time, Babe Ruth, was there, but not the greatest home-run hitter, Hank Aaron. The Marx Brothers were there, but not Orson Welles, Frank Capra, John Ford, or Steven Spielberg. Sarah Bernhardt was included, but not Leonard Bernstein. Rochester, New York, was on the list. Trenton, New Jersey, one of our most historic cities, was not. Hirsch included the Battle of the Bulge, which pleased my brother, who fought in it in 1944. But my uncle who died in the

Battle of the Coral Sea, in 1942, might have been disappointed to find that it didn't make the list.

To fill in the gaps, Hirsch has had to enlarge his list, so that there now exists a *Cultural Literacy Encyclopedia.* We may be sure that Hirsch will continue to expand his list until he reaches a point where a one-sentence directive will be all he needs to publish: "See the *Encyclopedia Americana* and *Webster's Third International.*"

It is, of course, an expected outcome of any education that students become acquainted with the important references of their culture. Even Rousseau, who would have asked his students to read only one book, *Robinson Crusoe* (so that they would learn how to survive in the wild), would probably have expected them to "pick up" the names and sayings and dates that made up the content of the educated conversation of their time. Nonetheless, Hirsch's proposal is inadequate for two reasons that reflect the inadequacies of Technopoly. The first, which I have discussed in chapter four, "The Improbable World," is that the present condition of technology-generated information is so long, varied, and dynamic that it is not possible to organize it into a coherent educational program. How do you include in the curriculum Rochester, New York, or Sarah Bernhardt or Babe Ruth? Or the Marx Brothers? Where does Ginger Rogers go? Does she get included in the syllabus under a unit titled "Fred Astaire's Dancing Partners"? (In which case, we must include Cyd Charisse and, if I am not mistaken, Winston Churchill's daughter, Sarah.) Hirsch's encyclopedic list is not a solution but a description of the problem of information glut. It is therefore essentially incoherent. But it also confuses a consequence of education with a purpose. Hirsch attempted to answer the question "What is an educated person?" He left unanswered the question "What is an education for?" Young men, for example, will learn how to make lay-up shots when they play basketball. To be able to make them is part of the definition of what good players are. But they do not play basketball for that purpose. There is usually a broader, deeper, and more meaningful reason for wanting to play—to assert their manhood, to please their fathers, to be acceptable to their peers, even for the sheer aesthetic pleasure of the game itself. What you have to do to be a success must be addressed only after you have found a reason to be successful. In Technopoly, this is very hard to do, and Hirsch simply sidestepped the question.

Not so Allan Bloom. In his book *The Closing of the American Mind,* he confronts the question by making a serious complaint against the academy. His complaint is that most American professors have lost their nerve. They have become moral relativists, incapable of providing their students with a clear understanding of what is right thought and proper behavior. Moreover, they are also intellectual relativists, refusing to defend their own culture and no longer committed to preserving and transmitting the best that has been thought and said.

Bloom's solution is that we go back to the basics of Western thought. He does not care if students know who Ginger Rogers and Groucho Marx are. He wants us to teach our students what Plato, Aristotle, Cicero, Saint Augustine, and other luminaries have had to say on the great ethical and epistemological questions. He believes that by acquainting themselves with great books our students will acquire a moral and intellectual foundation that will give meaning and texture to their lives. Though there is nothing especially original in this, Bloom is a serious education philosopher, which is to say, unlike Hirsch, he is a moralist who understands that Technopoly is a malevolent force requiring opposition. But he has not found many supporters.

Those who reject Bloom's idea have offered several arguments against it. The first is that such a purpose for education is elitist: the mass of students would not find the great story of

Western civilization inspiring, are too deeply alienated from the past to find it so, and would therefore have difficulty connecting the "best that has been thought and said" to their own struggles to find meaning in their lives. A second argument, coming from what is called a "leftist" perspective, is even more discouraging. In a sense, it offers a definition of what is meant by elitism. It asserts that the "story of Western civilization" is a partial, biased, and even oppressive one. It is not the story of blacks, American Indians, Hispanics, women, homosexuals—of any people who are not white heterosexual males of Judeo-Christian heritage. This claim denies that there is or can be a national culture, a narrative of organizing power and inspiring symbols which all citizens can identify with and draw sustenance from. If this is true, it means nothing less than that our national symbols have been drained of their power to unite, and that education must become a tribal affair; that is, each subculture must find its own story and symbols, and use them as the moral basis of education.

Standing somewhat apart from these arguments are, of course, religious educators, such as those in Catholic schools, who strive to maintain another traditional view—that learning is done for the greater glory of God and, more particularly, to prepare the young to embrace intelligently and gracefully the moral directives of the church. Whether or not such a purpose can be achieved in Technopoly is questionable, as many religious educators will acknowledge.

I will reserve for the next and final chapter my own view of the struggle to find a purpose for education in Technopoly. But here it must be said that the struggle itself is a sign that our repertoire of significant national, religious, and mythological symbols has been seriously drained of its potency. "We are living at a time," Irving Howe has written, "when all the once regnant world systems that have sustained (also distorted) Western intellectual life, from theologies to ideologies, are taken to be in severe collapse. This leads to a mood of skepticism, an agnosticism of judgment, sometimes a world-weary nihilism in which even the most conventional minds begin to question both distinctions of value and the value of distinctions."[2]

Into this void comes the Technopoly story, with its emphasis on progress without limits, rights without responsibilities, and technology without cost. The Technopoly story is without a moral center. It puts in its place efficiency, interest, and economic advance. It promises heaven on earth through the conveniences of technological progress. It casts aside all traditional narratives and symbols that suggest stability and orderliness, and tells, instead, of a life of skills, technical expertise, and the ecstasy of consumption. Its purpose is to produce functionaries for an ongoing Technopoly. It answers Bloom by saying that the story of Western civilization is irrelevant; it answers the political left by saying there is indeed a common culture whose name is Technopoly and whose key symbol is now the computer, toward which there must be neither irreverence nor blasphemy. It even answers Hirsch by saying that there are items on his list that, if thought about too deeply and taken too seriously, will interfere with the progress of technology.

I grant that it is somewhat unfair to expect educators, by themselves, to locate stories that would reaffirm our national culture. Such narratives must come to them, to some degree, from the political sphere. If our politics is symbolically impoverished, it is difficult to imagine how teachers can provide a weighty purpose to education. I am writing this chapter during the fourth week of the war against Iraq; the rhetoric accompanying the onset of the war is still fresh in mind. It began with the President's calling Americans to arms for the sake of their "life-style." This was followed by the Secretary of State's request that they fight to protect their jobs. Then came the appeal—at a late and convenient hour, as it were—to thwart the "naked aggression" of a little "Hitler." I do not say here that going to war was unjustified. My point is that, with the Cold War at an end, our political leaders now struggle, as never before, to find a vital narrative and accompanying symbols that would awaken a national spirit and a sense of resolve. The citizens themselves struggle as well. Having drained many of their traditional symbols of serious meaning, they resort, somewhat pitifully, to sporting yellow ribbons as a means of symbolizing their fealty to a cause. After the war, the yellow ribbons will fade from sight, but the question of who we are and what we represent will remain. Is it possible that the only symbol left to use will be an F-15 fighter plane guided by an advanced computer system?

11

The Loving Resistance Fighter

Anyone who practices the art of cultural criticism must endure being asked, What is the solution to the problems you describe? Critics almost never appreciate this question, since, in most cases, they are entirely satisfied with themselves for having posed the problems and, in any event, are rarely skilled in formulating practical suggestions about anything. This is why they became cultural critics.

The question comes forth nonetheless, and in three different voices. One is gentle and eager, as if to suggest that the critic knows the solutions but has merely forgotten to include them in the work itself. A second is threatening and judgmental, as if to suggest that the critic had no business bothering people in the first place unless there were some pretty good solutions at hand. And a third is wishful and encouraging, as if to suggest that it is well known that there are not always solutions to serious problems but if the critic will give it a little thought perhaps something constructive might come from the effort.

It is to this last way of posing the question that I should like to respond. I have indeed given the matter some thought, and this chapter is the result. Its simplicity will tell the reader that I am, like most other critics, armed less with solutions than with problems.

As I see it, a reasonable response (hardly a solution) to the problem of living in a developing Technopoly can be divided into two parts: what the individual can do irrespective of what the culture is doing; and what the culture can do irrespective of what any individual is doing. Beginning with the matter of individual response, I must say at once that I have no intention of providing a "how to" list in the manner of the "experts" I ridiculed in chapter five, on our "broken defenses." No one is an expert on how to live a life. I can, however, offer a Talmudic-like principle that seems to me an effective guide for those who wish to defend themselves against the worst effects of the American Technopoly. It is this: You must try to be a loving resistance fighter. That is the doctrine, as Hillel might say. Here is the commentary: By "loving," I mean that, in spite of the confusion, errors, and stupidities you see around you, you must always keep close to your heart the narratives and symbols that once made the United States the hope of the world and that may yet have enough vitality to do so again. You may find it helpful to remember that, when the Chinese students at Tiananmen Square gave expression to their impulse to democracy, they fashioned a papier-mâché model, for the whole world to see, of the Statue of Liberty. Not a statue of Karl Marx, not the Eiffel Tower, not Buckingham Palace. The Statue of Liberty. It is impossible to say how moved Americans were by this event. But one is compelled to ask, Is there an American soul so dead that it could not generate a murmur (if not a cheer) of satisfaction for this use of a once-resonant symbol? Is there an American soul so shrouded in the cynicism and malaise created by Technopoly's emptiness that it failed to be stirred by students reading aloud from the works of Thomas Jefferson in the streets of Prague in 1989? Americans may forget, but others do not, that American dissent and protest during the Vietnam War may be the only case in history where public opinion forced a government to change its foreign policy. Americans may forget, but others do not, that Americans invented the idea of public education for all citizens and have never abandoned it. And everyone knows, including Americans, that each day, to this hour, immigrants still come to America in hopes of finding relief from one kind of deprivation or another.

There are a hundred other things to remember that may help one to warm to the United States, including the fact that it has been, and perhaps always will be, a series of experiments that the world watches with wonder. Three such experiments are of particular importance. The first, undertaken toward the end of the eighteenth century, posed the question, Can a nation allow the greatest possible degree of political and religious freedom and still retain a sense of identity and purpose? Toward the middle of the nineteenth century, a second great experiment was undertaken, posing the question, Can a nation retain a sense of cohesion and community by allowing into it people from all over the world? And now comes the third—the great experiment of Technopoly—which poses the question, Can a nation preserve its history, originality, and humanity by submitting itself totally to the sovereignty of a technological thought-world?

Obviously, I do not think the answer to this question will be as satisfactory as the answers to the first two. But if there is an awareness of and resistance to the dangers of Technopoly, there is reason to hope that the United States may yet survive its Ozymandias-like hubris and technological promiscuity. Which brings me to the "resistance fighter" part of my principle. Those who resist the American Technopoly are people

who pay no attention to a poll unless they know what questions were asked, and why;

who refuse to accept efficiency as the pre-eminent goal of human relations;

who have freed themselves from the belief in the magical powers of numbers, do not regard calculation as an adequate substitute for judgment, or precision as a synonym for truth;

who refuse to allow psychology or any "social science" to pre-empt the language and thought of common sense;

who are, at least, suspicious of the idea of progress, and who do not confuse information with understanding;

who do not regard the aged as irrelevant;

who take seriously the meaning of family loyalty and honor, and who, when they "reach out and touch someone," expect that person to be in the same room;

who take the great narratives of religion seriously and who do not believe that science is the only system of thought capable of producing truth;

who know the difference between the sacred and the profane, and who do not wink at tradition for modernity's sake;

who admire technological ingenuity but do not think it represents the highest possible form of human achievement.

A resistance fighter understands that technology must never be accepted as part of the natural order of things, that every technology—from an IQ test to an automobile to a television set to a computer—is a product of a particular economic and political context and carries with it a program, an agenda, and a philosophy that may or may not be life-enhancing and that therefore require scrutiny, criticism, and control. In short, a technological resistance fighter maintains an epistemological and psychic distance from any technology, so that it always appears somewhat strange, never inevitable, never natural.

I can say no more than this, for each person must decide how to enact these ideas. But it is possible that one's education may help considerably not only in promoting the general conception of a resistance fighter but in helping the young to fashion their own ways of giving it expression. It is with education, then, that I will conclude this book. This is not to say that political action and social policy aren't useful in offering opposition to Technopoly. There are even now signs that Technopoly is understood as a problem to which laws and policies might serve as a response—in the environmental movement, in the contemplation of legal restrictions on computer technology, in a developing distrust of medical technology, in reactions against widespread testing, in various efforts to restore a sense of community cohesion. But in the United States, as Lawrence Cremin once remarked, whenever we need a revolution, we get a new curriculum. And so I shall propose one. I have done this before to something less than widespread acclamation.[1] But it is the best way I can think of for the culture to address the problem. School, to be sure, is a technology itself, but of a special kind in that, unlike most technologies, it is customarily and persistently scrutinized, criticized, and modified. It is America's principal instrument for correcting mistakes and for addressing problems that mystify and paralyze other social institutions.

In consideration of the disintegrative power of Technopoly, perhaps the most important contribution schools can make to the education of our youth is to give them a sense of coherence in their studies, a sense of purpose, meaning, and interconnectedness in what they learn. Modern secular education is failing not because it doesn't teach who Ginger Rogers, Norman Mailer, and a thousand other people are but because it has no moral, social, or intellectual center. There is no set of ideas or attitudes that permeates all parts of the curriculum. The curriculum is not, in fact, a "course of study" at all but a meaningless hodgepodge of subjects. It does not even put forward a clear vision of what constitutes an educated person, unless it is a person who possesses "skills." In other words, a technocrat's ideal—a person with no commitment and no point of view but with plenty of marketable skills.

Of course, we must not overestimate the capability of schools to provide coherence in the face of a culture in which almost all coherence seems to have disappeared. In our technicalized, present-centered information environment, it is not easy to locate a rationale for education, let alone impart one convincingly. It is obvious, for example, that the schools cannot restore religion to the center of the life of learning. With the exception of a few people, perhaps, no one would take seriously the idea that learning is for the greater glory of God. It is equally obvious that the knowledge explosion has blown apart the feasibility of such limited but coordinated curriculums as, for example, a Great Books program. Some people would have us stress love of country as a unifying principle in education. Experience has shown, however, that this invariably translates into love of government, and in practice becomes indistinguishable from what still is at the center of Soviet or Chinese education.

Some would put forward "emotional health" as the core of the curriculum. I refer here to a point of view sometimes called Rogerian, sometimes Maslovian, which values above all else the development of one's emotional life through the quest for one's "real self." Such an idea, of course, renders a curriculum irrelevant, since only "self-knowledge"—i.e., one's feelings—is considered worthwhile. Carl Rogers himself once wrote that anything that can be taught is probably either trivial or harmful, thus making any discussion of the schools unnecessary. But beyond this, the culture is already so heavy with the burden of the glorification of "self" that it would be redundant to have the schools stress it, even if it were possible.

One obviously treads on shaky ground in suggesting a plausible theme for a diverse, secularized population. Nonetheless, with all due apprehension, I would propose as a possibility the theme that animates Jacob Bronowski's *The Ascent of Man.* It is a book, and a philosophy, filled with optimism and suffused with the transcendent belief that humanity's destiny is the discovery of knowledge. Moreover, although Bronowski's emphasis is on science, he finds ample warrant to include the arts and humanities as part of our unending quest to gain a unified understanding of nature and our place in it.

Thus, to chart the ascent of man, which I will here call "the ascent of humanity," we must join art and science. But we must also join the past and the present, for the ascent of humanity is above all a continuous story. It is, in fact, a story of creation, although not quite the one that the fundamentalists fight so fiercely to defend. It is the story of humanity's creativeness in trying to conquer loneliness, ignorance, and disorder. And it certainly includes the development of various religious systems as a means of giving order and meaning to existence. In this context, it is inspiring to note that the Biblical version of creation, to the astonishment of everyone except possibly the fundamentalists, has turned out to be a near-perfect blend of artistic imagination and scientific intuition: the Big Bang theory of the creation of the universe, now widely accepted by cosmologists, confirms in essential details what the Bible proposes as having been the case "in the beginning."

In any event, the virtues of adopting the ascent of humanity as a scaffolding on which to build a curriculum are many and various, especially in our present situation. For one thing, with a few exceptions which I shall note, it does not require that we invent new subjects or discard old ones. The structure of the subject-matter curriculum that exists in most schools at present is entirely usable. For another, it is a theme that can begin in the earliest grades and extend through college in ever-deepening and -widening dimensions. Better still, it provides students with a point of view from which to understand the meaning of subjects, for each subject can be seen as a battleground of sorts, an arena in which fierce intellectual struggle has taken place and continues to take place. Each idea within a subject marks the place where someone fell and someone rose. Thus, the ascent of humanity is an optimistic story, not without its miseries but dominated by astonishing and repeated victories. From this point of view, the curriculum itself may be seen as a celebration of human intelligence and creativity, not a meaningless collection of diploma or college requirements.

Best of all, the theme of the ascent of humanity gives us a nontechnical, noncommercial definition of education. It is a definition drawn from an honorable humanistic tradition and reflects a concept of the purposes of academic life that goes counter to the biases of the technocrats. I am referring to the idea that to become educated means to become aware of the origins and growth of knowledge and knowledge systems; to be familiar with the intellectual and creative processes by which the best that has been thought and said has been produced; to learn how to participate, even if as a listener, in what Robert Maynard Hutchins once called The Great Conversation, which is merely a different metaphor for what is meant by the ascent of humanity. You will note that such a definition is not child-centered, not training-centered, not skill-centered, not even problem-centered. It is idea-centered and coherence-centered. It is also otherworldly, inasmuch as it does not assume that what one learns in school must be directly and urgently related to a problem of today. In other words, it is an education that stresses history, the scientific mode of thinking, the disciplined use of language, a wide-ranging knowledge of the arts and religion, and the continuity of human enterprise. It is education as an excellent corrective to the antihistorical, information-saturated, technology-loving character of Technopoly.

Let us consider history first, for it is in some ways the central discipline in all this. It is hardly necessary for me to argue here that, as Cicero put it, "To remain ignorant of things that happened before you were born is to remain a child." It is enough to say that history is our most potent intellectual means of achieving a "raised consciousness." But there are some points about history and its teaching that require stressing, since they are usually ignored by our schools. The first is that history is not merely one subject among many that may be taught; *every* subject has a history, including biology, physics, mathematics, literature, music, and art. I would propose here that every teacher must be a history teacher. To teach, for example, what we know about biology today without also teaching what we once knew, or thought we knew, is to reduce knowledge to a mere consumer product. It is to deprive students of a sense of the meaning of what we know, and of how we know. To teach about the atom without Democritus, to teach about electricity without Faraday, to teach about political science without Aristotle or Machiavelli, to teach about music without Haydn, is to refuse our students access to The Great Conversation. It is to deny them knowledge of their roots, about which no other social institution is at present concerned. For to know about your roots is not merely to know where your grandfather came from and what he had to endure. It is also to know where your ideas come from and why you happen to believe them; to know where your moral and aesthetic sensibilities come from. It is to know where your world, not just your family, comes from. To complete the presentation of Cicero's thought, begun above: "What is a human life worth unless it is incorporated into the lives of one's ancestors and set in an historical context?" By "ancestors" Cicero did not mean your mother's aunt.

Thus, I would recommend that every subject be taught *as* history. In this way, children, even in the earliest grades, can begin to understand, as they now do not, that knowledge is not a fixed thing but a stage in human development, with a past and a future. To return for a moment to theories of creation, we want to be able to show how an idea conceived almost four thousand years ago has traveled not only in time but in meaning, from science to religious metaphor to science again. What a lovely and profound coherence there is in the connection between the wondrous speculations in an ancient Hebrew desert tent and the equally wondrous speculations in a modern MIT classroom! What I am trying to say is that the history of subjects teaches connections; it teaches that the world is not created anew each day, that everyone stands on someone else's shoulders.

I am well aware that this approach to subjects would be difficult to use. There are, at present, few texts that would help very much, and teachers have not, in any case, been prepared to know about knowledge in this way. Moreover, there is the added difficulty of our learning how to do this for children of different ages. But that it needs to be done is, in my opinion, beyond question.

The teaching of subjects as studies in historical continuities is not intended to make history as a special subject irrelevant. If every subject is taught with a historical dimension, the history teacher will be free to teach what histories are: hypotheses and theories about why change occurs. In one sense, there is no such thing as "history," for every historian from Thucydides to Toynbee has known that his stories must be told from a special point of view that will reflect his particular theory of social development. And historians also know that they write histories for some particular purpose—more often than not, either to glorify or to condemn the present. There is no definitive history of anything; there are only histories, human inventions which do not give us *the* answer, but give us only those answers called forth by the questions that have been asked.

Historians know all of this—it is a commonplace idea among them. Yet it is kept a secret from our youth. Their ignorance of it prevents them from understanding how "history" can change and why the Russians, Chinese, American Indians, and virtually everyone else see historical events differently than the authors of history schoolbooks. The task of the history teacher, then, is to become a "histories teacher." This does not mean that some particular version of the American, European, or Asian past should remain untold. A student who does not know at least one history is in no position to evaluate others. But it does mean that a histories teacher will be concerned, at all times, to show how histories are themselves products of culture; how any history is a mirror of the conceits and even metaphysical biases of the culture that produced it; how the religion, politics, geography, and economy of a people lead them to re-create their past along certain lines. The histories teacher must clarify for students the meaning of "objectivity" and "events," must show what a "point of view" and a "theory" are, must provide some sense of how histories may be evaluated.

It will be objected that this idea—history as comparative history—is too abstract for students to grasp. But that is one of the several reasons why comparative history should be taught. To teach the past simply as a chronicle of indisputable, fragmented, and concrete events is to replicate the bias of Technopoly, which largely denies our youth access to concepts and theories, and to provide them only with a stream of meaningless events. That is why the controversies that develop around what events ought to be included in the "history" curriculum have a somewhat hollow ring to them. Some people urge, for example, that the Holocaust, or Stalin's bloodbaths, or the trail of Indian tears be taught in school. I agree that our students should know about such things, but we must still address the question, What is it that we want them to "know" about these events? Are they to be explained as the "maniac" theory of history? Are they to be understood as illustrations of the "banality of evil" or the "law of survival"? Are they manifestations of the universal force of economic greed? Are they examples of the workings of human nature?

Whatever events may be included in the study of the past, the worst thing we can do is to present them devoid of the coherence that a theory or theories can provide—that is to say, as meaningless. This, we can be sure, Technopoly does daily. The histories teacher must go far beyond the "event" level into the realm of concepts, theories, hypotheses, comparisons, deductions, evaluations. The idea is to raise the level of abstraction at which "history" is taught. This idea would apply to all subjects, including science.

From the point of view of the ascent of humanity, the scientific enterprise is one of our most glorious achievements. On humanity's Judgment Day we can be expected to speak almost at once of our science. I have already stressed the importance of teaching the history of science in every science course, but this is no more important than teaching its "philosophy." I mention this with some sense of despair. More than half the high schools in the United States do not even offer one course in physics. And at a rough guess, I would estimate that in 90 percent of the schools chemistry is still taught as if students were being trained to be druggists. To suggest, therefore, that science is an exercise in human imagination, that it is something quite different from technology, that there are "philosophies" of science, and that all of this ought to form part of a scientific education, is to step out of the mainstream. But I believe it nonetheless.

Would it be an exaggeration to say that not one student in fifty knows what "induction" means? Or knows what a scientific theory is? Or a scientific model? Or knows what are the optimum conditions of a valid scientific experiment? Or has ever considered the question of what scientific truth is? In *The Identity of Man* Bronowski says the following: "This is the paradox of imagination in science, that it has for its aim the impoverishment of imagination. By that outrageous phrase, I mean that the highest flight of scientific imagination is to weed out the proliferation of new ideas. In science, the grand view is a miserly view, and a rich model of the universe is one which is as poor as possible in hypotheses."

Is there one student in a hundred who can make any sense out of this statement? Though the phrase "impoverishment of imagination" may be outrageous, there is nothing startling or even unusual about the idea contained in this quotation. Every practicing scientist understands what Bronowski is saying. Yet it is kept a secret from our students. It should be revealed. In addition to having each science course include a serious historical dimension, I would propose that every school—elementary through college—offer and require a course in the philosophy of science. Such a course should consider the language of science, the nature of scientific proof, the source of scientific hypotheses, the role of imagination, the conditions of experimentation, and especially the value of error and disproof. If I am not mistaken, many people still believe that what makes a statement scientific is that it can be verified. In fact, exactly the opposite is the case: What separates scientific statements from nonscientific statements is that the former can be subjected to the test of falsifiability. What makes science possible is not our ability to recognize "truth" but our ability to recognize falsehood.

What such a course would try to get at is the notion that science is not pharmacy or technology or magic tricks but a special way of employing human intelligence. It would be important for students to learn that one becomes scientific not by donning a white coat (which is what television teaches) but by practicing a set of canons of thought, many of which have to do with the disciplined use of language. Science involves a method of employing language that is accessible to everyone. The ascent of humanity has rested largely on that.

On the subject of the disciplined use of language, I should like to propose that, in addition to courses in the philosophy of science, every school—again, from elementary school through college—offer a course in semantics—in the processes by which people make meaning. In this connection I must note the gloomy fact that English teachers have been consistently obtuse in their approach to this subject—which is to say, they have largely ignored it. This has always been difficult for me to understand, since English teachers claim to be concerned with teaching reading and writing. But if they do not teach anything about the relationship of language to reality—which is what semantics studies—I cannot imagine how they expect reading and writing to improve.

Every teacher ought to be a semantics teacher, since it is not possible to separate language from what we call knowledge. Like history, semantics is an interdisciplinary subject: it is necessary to know something about it in order to understand any subject. But it would be extremely useful to the growth of their intelligence if our youth had available a special course in which fundamental principles of language were identified and explained. Such a course would deal not only with the various uses of language but with the relationship between things and words, symbols and signs, factual statements and judgments, and grammar and thought. Especially for young students, the course ought to emphasize the kinds of semantic errors that are common to all of us, and that are avoidable through awareness and discipline—the use of either-or categories, misunderstanding of levels of abstraction, confusion of words with things, sloganeering, and self-reflexiveness.

Of all the disciplines that might be included in the curriculum, semantics is certainly among the most "basic." Because it deals with the processes by which we make and interpret meaning, it has great potential to affect the deepest levels of student intelligence. And yet semantics is rarely mentioned when "back to the basics" is proposed. Why? My guess is that it cuts too deep. To adapt George Orwell, many subjects are basic but some are more basic than others. Such subjects have the capability of generating critical thought and of giving students access to questions that get to the heart of the matter. This is not what "back to the basics" advocates usually have in mind. They want language technicians: people who can follow instructions, write reports clearly, spell correctly. There is certainly ample evidence that the study of semantics will improve the writing and reading of students. But it invariably does more. It helps students to reflect on the sense and truth of what they are writing and of what they are asked to read. It teaches them to discover the underlying assumptions of what they are told. It emphasizes the manifold ways in which language can distort reality. It assists students in becoming what Charles Weingartner and I once called "crap-detectors." Students who have a firm grounding in semantics are therefore apt to find it difficult to take reading tests. A reading test does not invite one to ask whether or not what is written is true. Or, if it is true, what it has to do with anything. The study of semantics insists upon these questions. But "back to the basics" advocates don't require education to be *that* basic. Which is why they usually do not include literature, music, and art as part of their agenda either. But of course, in using the ascent of humanity as a theme, we would of necessity elevate these subjects to prominence.

The most obvious reason for such prominence is that their subject matter contains the best evidence we have of the unity and continuity of human experience and feeling. And that is why I would propose that, in our teaching of the humanities, we should emphasize the enduring creations of the past. The schools should stay as far from contemporary works as possible. Because of the nature of the communications industry, our students have continuous access to the popular arts of their own times—its music, rhetoric, design, literature, architecture. Their knowledge of the form and content of these arts is by no means satisfactory. But their ignorance of the form and content of the art of the past is cavernous. This is one good reason for emphasizing the art of the past. Another is that there is no subject better suited to freeing us from the tyranny of the present than the historical study of art. Painting, for example, is more than three times as old as writing, and contains in its changing styles and themes a fifteen-thousand-year-old record of the ascent of humanity.

In saying this, I do not mean to subsume art under the heading of archeology, although I should certainly recommend that the history of art forms be given a serious place in the curriculum. But art is much more than a historical artifact. To have meaning for us, it must connect with those levels of feeling that are in fact not expressible in discursive language. The question therefore arises whether it is possible for students of today to relate, through feeling, to the painting, architecture, music, sculpture, or literature of the past. The answer, I believe, is: only with the greatest difficulty. They, and many of us, have an aesthetic sensibility of a different order from what is required to be inspired, let alone entertained, by a Shakespeare sonnet, a Haydn symphony, or a Hals painting. To oversimplify the matter, a young man who believes Madonna to have reached the highest pinnacle of musical expression lacks the sensibility to distinguish between the ascent and descent of humanity. But it is not my intention here to blacken the reputation of popular culture. The point I want to make is that the products of the popular arts are amply provided by the culture itself. The schools must make available the products of classical art forms precisely because they are not so available and because they demand a different order of sensibility and response. In our present circumstances, there is no excuse for schools to sponsor rock concerts when students have not heard the music of Mozart, Beethoven, Bach, or Chopin. Or for students to have graduated from high school without having read, for example, Shakespeare, Cervantes, Milton, Keats, Dickens, Whitman, Twain, Melville, or Poe. Or for students not to have seen at least a photograph of paintings by Goya, El Greco, David. It is not to the point that many of these composers, writers, and painters were in their own times popular artists. What is to the point is that they spoke, when they did, in a language and from a point of view different from our own and yet continuous with our own. These artists are relevant not only because they established the standards with which civilized people approach the arts. They are relevant because the culture tries to mute their voices and render their standards invisible.

It is highly likely that students, immersed in today's popular arts, will find such an emphasis as I suggest tedious and even painful. This fact will, in turn, be painful to teachers, who, naturally enough, prefer to teach that which will arouse an immediate and enthusiastic response. But our youth must be shown that not all worthwhile things are instantly accessible and that there are levels of sensibility unknown to them. Above all, they must be shown humanity's artistic roots. And that task, in our own times, falls inescapably to the schools.

On the matter of roots, I want to end my proposal by including two subjects indispensable to any understanding of where we have come from. The first is the history of technology, which as much as science and art provides part of the story of humanity's confrontation with nature and indeed with our own limitations. It is important for students to be shown, for example, the connection between the invention of eyeglasses in the thirteenth century and experiments in gene-splicing in the twentieth: that in both cases we reject the proposition that anatomy is destiny, and through technology define our own destiny. In brief, we need students who will understand the relationships between our technics and our social and psychic worlds, so that they may begin informed conversations about where technology is taking us and how.

The second subject is, of course, religion, with which so much painting, music, technology, architecture, literature, and science are intertwined. Specifically, I want to propose that the curriculum include a course in comparative religion. Such a course would deal with religion as an expression of humanity's creativeness, as a total, integrated response to fundamental questions about the meaning of existence. The course would be descriptive, promoting no particular religion but illuminating the metaphors, the literature, the art, the ritual of religious expression itself. I am aware of the difficulties such a course would face, not the least of which is the belief that the schools and religion must on no account touch each other. But I do not see how we can claim to be educating our youth if we do not ask them to consider how different people of different times and places have tried to achieve a sense of transcendence. No education can neglect such sacred texts as Genesis, the New Testament, the Koran, the Bhagavad-Gita. Each of them embodies a style and a world-view that tell as much about the ascent of humanity as any book ever written. To these books I would add the *Communist Manifesto*, since I think it reasonable to classify this as a sacred text, embodying religious principles to which millions of people have so recently been devoted.

To summarize: I am proposing, as a beginning, a curriculum in which all subjects are presented as a stage in humanity's historical development; in which the philosophies of science, of history, of language, of technology, and of religion are taught; and in which there is a strong emphasis on classical forms of artistic expression. This is a curriculum that goes "back to the basics," but not quite in the way the technocrats mean it. And it is most certainly in opposition to the spirit of Technopoly. I have no illusion that such an education program can bring a halt to the thrust of a technological thought-world. But perhaps it will help to begin and sustain a serious conversation that will allow us to distance ourselves from that thought-world, and then criticize and modify it. Which is the hope of my book as well.

Notes

ONE

1. Plato, p. 96.

2. Freud, pp. 38–39.

3. This fact is documented in Keith Hoskin's "The Examination, Disciplinary Power and Rational Schooling," in *History of Education,* vol. VIII, no. 2 (1979), pp. 135–46. Professor Hoskin provides the following story about Farish: Farish was a professor of engineering at Cambridge and designed and installed a movable partition wall in his Cambridge home. The wall moved on pulleys between downstairs and upstairs. One night, while working late downstairs and feeling cold, Farish pulled down the partition. This is not much of a story, and history fails to disclose what happened next. All of which shows how little is known of William Farish.

4. For a detailed exposition of Mumford's position on the impact of the mechanical clock, see his *Technics and Civilization.*

TWO

1. Marx, p. 150.

2. Perhaps another term for a tool-using culture is "third-world country," although vast parts of China may be included as tool-using.

3. For a detailed analysis of medieval technology, see Jean Gimpel's *The Medieval Machine.*

4. Quoted in Muller, p. 30.

5. See his *Medieval Technology and Social Change.*

6. De Vries' findings are recounted by Alvin Toffler in his article "Value Impact Forecaster: A Profession of the Future," in Baier and Rescher's book *Values and the Future: The Impact of Technological Change on American Values* (New York: Free Press, 1969), p. 3.

THREE

1. Giedion, p. 40.

2. The best account of the history of utopias may be found in Segal.

3. See David Linton's "Luddism Reconsidered" in *Etcetera,* Spring 1985, pp. 32–36.

4. Tocqueville, p. 404.

FOUR

1. For a detailed examination of the impact of the printing press on Western culture, see Eisenstein.

2. See Postman's *Amusing Ourselves to Death* for a more full-bodied treatment of the telegraph.

FIVE

1. An emphatic exception among those sociologists who have written on this subject is Arnold Gehlen. See his *Man in the Age of Technology.*

2. Though this term is by no means original with E. D. Hirsch, Jr., its current popularity is attributable to Hirsch's book *Cultural Literacy.*

3. This poignant phrase is also the title of one of Lasch's most important books.

4. James Beniger, *The Control Revolution,* p. 13. As I have already noted, Beniger's book is the best source for an understanding of the technical means of eliminating—i.e., controlling—information.

5. Tocqueville, p. 262.

6. Lewis, p. x.

7. See Arendt.

SIX

1. I am not sure whether the company still exists, but by way of proving that it at least once did, here is the address of the HAGOTH Corporation as I once knew it: 85 NW Alder Place, Department C, Issaquah, Washington 98027.

2. All these facts and more may be found in Payer, or in Inlander et al.

3. Reiser, p. 160.

4. Ibid., p. 161.

5. Payer, p. 127.

6. Quoted in ibid.

7. For a fascinating account of Laënnec's invention, see Reiser.

8. Ibid., p. 38.

9. Ibid., p. 230.

10. Horowitz, p. 31.

11. Ibid., p. 80.

12. Cited in Inlander et al., p. 106.

13. Cited in ibid., p. 113.

SEVEN

1. *New York Times,* August 7, 1990, sect. C, p. 1.

2. *Personal Computing,* June 29, 1990, p. 36.

3. *New York Times,* November 24, 1989.

4. *Publishers Weekly,* March 2, 1990, p. 26.

5. *Bottom Line,* July 15, 1989, p. 5.

6. For a concise and readable review of the development of the computer, I would recommend Arno Penzias' *Ideas and Information: Managing in a High-Tech World.*

7. Quoted in Hunt, p. 318.

8. Searle, p. 30.

9. See Gozzi, pp. 177–80.

10. See Milgram.

11. Weizenbaum, p. 32.

12. The March 1991 issue of *The Sun* reports that Lance Smith, who is two years old, is called "the Mozart of video games," mainly because he gets astronomical scores on one of Nintendo's games. This is as close to approaching the artistry of Mozart as computers can get.

13. See J. D. Bolter's 1991 book, *Writing Space: The Computer, Hypertext and the History of Writing* (Hillsdale, N.J.: Lawrence Erlbaum Associates).

14. *Science Digest,* June 1984.

15. Both men are quoted in the Raleigh, North Carolina, *News and Observer,* Sunday, August 13, 1989.

16. Katsch, p. 44.

EIGHT

1. Cited in Gould, p. 75. I am indebted to Gould's wonderful book for providing a concise history of the search to quantify intelligence.

2. *The National Elementary Principal,* March/April 1975.

3. Weizenbaum, p. 203.

4. The occasion, in the spring of 1990, was a retreat outside of Washington, D.C. The group of twenty-three Democratic congressmen was led by Richard Gephardt.

5. I have, of course, made up these ridiculous statistics. The point is, it doesn't matter.

6. See the preceding note.

7. An interesting example of the tyranny of statistics is in the decision made by the College Board (on November 1, 1990) that its Scholastic Aptitude Test will not include asking students to write an essay. To deter-

mine the student's ability to write, the SAT will continue to use a multiple-choice test that measures one's ability to memorize rules of grammar, spelling, and punctuation. It would seem reasonable—wouldn't it?—that the best way to find out how well someone writes is to ask him or her to write something. But in Technopoly reason is a strange and wondrous thing. For a documentation of all of this, see the January 16, 1991, issue of *The Chronicle of Higher Education.*

8. See Keith W. Hoskin and Richard H. Macve, "The Genesis of Accountability: The West Point Connections," in *Accounting Organizations and Society,* vol. 13, no. 1 (1988), pp. 37–73. I am especially indebted to these scholars for their account of the development of modern systems of management.

NINE

1. Cited in Hayek, p. 201. I am indebted to Hayek's book for his history of the Ecole Polytechnique.

2. Ibid., p. 21.

3. Myrdal, p. 6.

4. I have borrowed much of the material dealing with the distinctions between natural science and social research from my own essay "Social Science as Moral Theology," in *Conscientious Objections.*

TEN

1. Although in some ways Boorstin's book is dated, to him and his book go credit for calling early attention to the effects of an image society.

2. *The New Republic,* February 18, 1991, p. 42.

ELEVEN

1. What follows is a version of a proposal I have made several times before. A somewhat fuller version appears in my *Teaching as a Conserving Activity.*

Bibliography

Al-Hibri, A., and Hickman, L. (eds.). *Technology and Human Affairs.* London: The C. V. Mosby Company, 1981.

Arendt, H. *Eichmann in Jerusalem: A Report on the Banality of Evil.* New York: Penguin Books, 1977.

Bellah, R. N.; Madsen, R.; Sullivan, W. H.; Swidler, A.; and Tipton, S. M. *Habits of the Heart: Individualism and Commitment in American Life.* Berkeley: University of California Press, 1985.

Beniger, J. R. *The Control Revolution: Technological and Economic Origins of the Information Society.* Cambridge, Mass., and London: Harvard University Press, 1986.

Bolter, J. D. *Turing's Man: Western Culture in the Computer Age.* Chapel Hill: The University of North Carolina Press, 1984.

Bury, J. B. *The Idea of Progress: An Inquiry into its Origin and Growth.* New York: Dover Publications, Inc., 1932.

Callahan, R. E. *Education and the Cult of Efficiency: A Study of the Social Forces That Have Shaped the Administration of the Public Schools.* Chicago: The University of Chicago Press, 1962.

Christians, C. G., and Van Hook, J. M. (eds.). *Jacques Ellul: Interpretive Essays.* Chicago: University of Illinois Press, 1981.

Eisenstein, E. *The Printing Revolution in Early Modern Europe.* Cambridge, Mass.: Cambridge University Press, 1983.

Ellul, J. *The Technological Society.* New York: Alfred A. Knopf, 1964.

Ellul, J. *The Betrayal of the West.* New York: The Seabury Press, 1978.

Farrington, B. *Francis Bacon: Philosopher of Industrial Science.* New York: Henry Schuman, Inc., 1949.

Freud, S. *Civilization and Its Discontents.* New York: W. W. Norton & Co., 1961.

Gehlen, A. *Man in the Age of Technology.* New York: Columbia University Press, 1980.

Giedion, S. *Mechanization Takes Command: A Contribution to Anonymous History.* New York: W. W. Norton & Co., 1948.

Gimpel, J. *The Medieval Machine: The Industrial Revolution of the Middle Ages.* New York: Holt, Rinehart & Winston, 1976.

Gould, S. J. *The Mismeasure of Man.* New York: W. W. Norton & Co., 1981.

Gozzi, R., Jr. "The Computer 'Virus' as Metaphor," in *Etcetera: A Review of General Semantics,* vol. 47, no. 2 (Summer 1990).

Hayek, F. H. *The Counter-Revolution of Science: Studies on the Abuse of Reason.* Indianapolis: Liberty Press, 1952.

Hirsch, E. D., Jr. *Cultural Literacy: What Every American Needs to Know.* Boston: Houghton Mifflin Co., 1987.

Hodges, A. *Alan Turing: The Enigma.* New York: Simon & Schuster, 1983.

Hoffer, E. *The Ordeal of Change.* New York: Harper & Row, 1952.

Horowitz, L. C., M.D. *Taking Charge of Your Medical Fate.* New York: Random House, 1988.

Hunt, M. *The Universe Within: A New Science Explores the Mind.* New York: Simon & Schuster, 1982.

Hutchins, R. M. *The Higher Learning in America.* New Haven: Yale University Press, 1936.

Inlander, C. B.; Levin, L. S.; and Weiner, E. *Medicine on Trial: The Appalling Story of Medical Ineptitude and the Arrogance that Overlooks It.* New York: Pantheon Books, 1988.

Katsch, M. E. *The Electronic Media and the Transformation of Law.* New York and Oxford: Oxford University Press, 1989.

Koestler, A. *The Sleepwalkers.* New York: The Macmillan Company, 1968.

Lasch, C. *Haven in a Heartless World: The Family Besieged.* New York: Basic Books, Inc., 1975.

Lewis, C. S. *The Screwtape Letters.* New York: Macmillan, 1943.

Logan, R. K. *The Alphabet Effect: The Impact of the Phonetic Alphabet on the Development of Western Civilization.* New York: St. Martin's Press, 1986.

Luke, C. *Pedagogy, Printing, and Protestantism.* Albany: State University of New York Press, 1989.

Marx, K., and Engels, F. *The German Ideology.* New York: International Publishers, 1972.

Milgram, S. *Obedience to Authority: An Experimental View.* New York: Harper & Row, 1974.

Muller, H. J. *The Children of Frankenstein: A Primer on Modern Technology and Human Values.* Bloomington and London: Indiana University Press, 1970.

Mumford, L. *Technics and Civilization.* New York: Harcourt Brace Jovanovich, 1963.

Myrdal, G. *Objectivity in Social Research.* New York: Pantheon Books, 1969.

Papert, S. *Mindstorms: Children, Computers, and Powerful Ideas.* New York: Basic Books, Inc., 1980.

Payer, L. *Medicine and Culture: Varieties of Treatment in the United States, England, West Germany, and France.* New York: Penguin Books, 1988.

Penzias, A. *Ideas and Information: Managing in a High-Tech World.* New York and London: W. W. Norton & Co., 1989.

Plato. *Phaedrus and Letters VII and VIII.* New York: Penguin Books, 1973.

Postman, N. *Amusing Ourselves to Death: Public Discourse in the Age of Show Business.* New York: Penguin Books, 1985.

Read, H. *To Hell with Culture and Other Essays on Art and Society.* New York: Schocken Books, 1963.

Reiser, S. J. *Medicine and the Reign of Technology.* Cambridge, London, New York and Melbourne: Cambridge University Press, 1978.

Rifkin, J. *Time Wars: The Primary Conflict in Human History.* New York: Henry Holt and Company, 1987.

Schumacher, E. F. *Small Is Beautiful: Economics As If People Mattered.* New York, Hagerstown, San Francisco and London: Harper & Row.

Schumacher, E. F. *A Guide for the Perplexed.* New York: Viking Penguin, Inc., 1977.

Searle, J. *Minds, Brains and Science.* Cambridge, Mass.: Harvard University Press, 1984.

Segal, H. P. *Technological Utopianism in American Culture.* Chicago: The University of Chicago Press, 1985.

Snow, C. P. *The Two Cultures and the Scientific Revolution.* New York: Cambridge University Press, 1959.

Sturt, M. *Francis Bacon.* New York: William Morrow & Company, 1932.

Szasz, T. *Anti-Freud: Karl Kraus's Criticism of Psychoanalysis and Psychiatry.* Syracuse: Syracuse University Press, 1976.

Tocqueville, A. de. *Democracy in America.* New York: Anchor Books (Doubleday & Co., Inc.), 1969.

Usher, A. P. *History of Mechanical Inventions.* New York: Dover Publications, Inc., 1929.

Weingartner, C. "Educational Research: The Romance of Quantification," *Etcetera: A Review of General Semantics,* vol. 39, no. 2 (Summer 1982).

Weizenbaum, J. *Computer Power and Human Reason: From Judgment to Calculation.* San Francisco: W. H. Freeman and Company, 1976.

White, L., Jr. *Medieval Technology and Social Change.* London: Oxford University Press, 1962.

Whitehead, A. N. *The Aims of Education and Other Essays.* New York: The Free Press, 1929.

Whitrow, G. J. *Time in History: The Evolution of Our General Awareness of Time and Temporal Perspective.* Oxford and New York: Oxford University Press, 1988.

Index